W9-AWD-527

Physlet® Quantum Physics
An Interactive Introduction

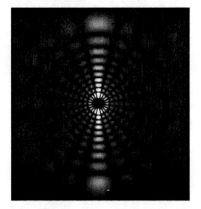

Mario Belloni
Wolfgang Christian
Anne J. Cox

PEARSON
Prentice
Hall

Upper Saddle River, New Jersey 07458

Library of Congress Cataloging-in-Publication Data

Belloni, Mario.
Physlet® quantum physics: an interactive introduction / Mario Belloni,
Wolfgang Christian, Anne J. Cox.
p. cm.– (PH series in educational innovation)
Includes bibliographical references.
ISBN 0-13-101970-8
1. Quantum theory–Computer-assisted instruction–Textbooks. 2. Quantum theory–Problems, exercises, etc. I. Christian, Wolfgang. II. Cox, Anne J. III. Title. IV. Prentice Hall series in educational innovation.

QC174.12.B445 2006
530.12'078'5–dc22   2005015232

Associate Editor: *Christian Botting*
Senior Editor: *Erik Fahlgren*
Executive Managing Editor: *Kathleen Schiaparelli*
Production Editor: *Debra Wechsler*
Assistant Managing Editor, Science Media: *Nicole Jackson*
Media Production Editor: *Tyler Suydam*
Manufacturing Buyer: *Alan Fischer*
Manufacturing Manager: *Alexis Heydt-Long*
Senior Marketing Manager: *Shari Meffert*
Art Director: *Jayne Conte*
Editorial Assistant: *Jessica Berta*
Marketing Assistant: *Laura Rath*
Production Assistant: *Nancy Bauer*

© 2006 Pearson Education, Inc.
Pearson Prentice Hall
Pearson Education, Inc.
Upper Saddle River, New Jersey 07458

All rights reserved. No part of this book or the accompanying CD may be reproduced or transmitted in any form without permission in writing from the publisher in English and Mario Belloni, Wolfgang Christian, and Anne J. Cox in all other languages. Exercises using Open Source Physics (OSP) applets are copyright Mario Belloni, Wolfgang Christian, and Anne J. Cox in all languages and are clearly marked. Physlets®, that is, the applets themselves, are copyrighted and trademarked by Wolfgang Christian. Physlets may be used and freely distributed for non-profit, educational purposes without requesting permission as outlined in the conditions of use.

Pearson Prentice Hall™ is a trademark of Pearson Education, Inc.

Printed in the United States of America
10 9 8 7

ISBN 0-13-101970-8

Pearson Education LTD., *London*
Pearson Education Australia PTY, Limited, *Sydney*
Pearson Education Singapore, Pte. Ltd
Pearson Education North Asia Ltd, *Hong Kong*
Pearson Education Canada, Ltd., *Toronto*
Pearson Educacion de Mexico, S.A. de C.V.
Pearson Education - Japan, *Tokyo*
Pearson Education Malaysia, Pte. Ltd

ei Prentice Hall Series in Educational Innovation

Lecture Tutorials for Introductory Astronomy
Jeffrey P. Adams
Edward E. Prather
Timothy F. Slater
Jack Dostal
The Conceptual Astronomy and Physics Education Research (CAPER) Team

Physlet® Physics: Interactive Illustrations, Explorations and Problems for Introductory Physics
Wolfgang Christian
Mario Belloni

Physlets®: Teaching Physics with Interactive Curricular Material
Wolfgang Christian
Mario Belloni

Peer Instruction for Astronomy
Paul J. Green

E&M TIPERs: Electricity & Magnetism Tasks Inspired by Physics Education Research
Curt J. Hieggelke
David P. Maloney
Stephen E. Kanim
Thomas L. O'Kuma

Peer Instruction: A User's Manual
Eric Mazur

Tutorials in Introductory Physics and Homework Package
Lillian C. McDermott
Peter S. Shaffer
The Physics Education Group, University of Washington

Just-In-Time Teaching: Blending Active Learning with Web Technology
Gregor M. Novak
Evelyn T. Patterson
Andrew D. Gavrin
Wolfgang Christian

Ranking Task Exercises in Physics: Student Edition
Thomas L. O'Kuma
David P. Maloney
Curt J. Hieggelke

Learner-Centered Astronomy Teaching: Strategies for ASTRO 101
Timothy F. Slater
Jeffrey P. Adams

About the Authors

Mario Belloni is an associate professor of physics at Davidson College. He received his B.A. in physics and economics from the University of California, Berkeley and his Ph.D. in physics from the University of Connecticut at Storrs. His research interests are in the areas of theoretical physics and interactive curricular material development. He is the co-author of *Fislets*, *Physlet Physics* (Prentice Hall 2004), and *Physlets: Teaching Physics with Interactive Curricular Material* (Prentice Hall 2001). He is the past Chair of the Committee on Educational Technologies of the AAPT, is the North Carolina Section Representative of the AAPT, and is a member of the ComPADRE Quantum Physics Editorial Board.

mabelloni@davidson.edu

Wolfgang Christian is the Herman Brown Professor of Physics at Davidson College where he has taught since 1983. He received his B.S. and Ph.D. in physics from North Carolina State University at Raleigh. He is the co-author of *Fislets*, *Physlet Physics*, *Physlets*, *Just-in-Time Teaching*, and *Volume 9 of the Computational Physics Upper Level Software: Waves and Optics Simulations*. He is past Chair of the American Physical Society Forum on Education. He has received a Distinguished Service Citation from the American Association of Physics Teachers. His research interests are in the areas of computational physics and instructional software design.

wochristian@davidson.edu

Anne J. Cox, associate professor of physics at Eckerd College, has taught at Eckerd for eight years. She has a B.S. in physics from Rhodes College and a Ph.D. in physics from the University of Virginia. In 2004 she was awarded Eckerd's Staub Distinguished Teacher of the Year Award. Her current research interests are curriculum development and pedagogical strategies to enhance student learning using technology. She is a contributing author of *Physlet Physics*. She is the current President of the Florida Section of the AAPT and a member of the Committee on Education Technologies of the AAPT.

coxaj@eckerd.edu

Contents

Preface   xiii

Java Tests and Browser Requirements   xvii

1 Introduction to Physlets   1
Introduction 1
1.1   Static Images Versus Physlet Animations 1
1.2   Animations, Units, and Measurement 3
1.3   Exploring the Input of Data: Numbers 5
1.4   Exploring the Input of Data: Formulas 6
1.5   Getting Data Out 7
1.6   Exploring the Input of Data: Complex Expressions 8
Problems 9

I   Special Relativity   11

2 Space and Time in Special Relativity   12
Introduction 12
2.1   Synchronizing Clocks 12
2.2   Exploring Synchronizing Clocks by Viewing 13
2.3   Simultaneity 14
2.4   Light Clocks, Time Dilation, and Length Contraction 15
2.5   Understanding Spacetime Diagrams 17
2.6   Exploring Worldlines by Matching 18
2.7   Exploring the Pole and Barn Paradox 19
2.8   Exploring the Twin Paradox 20
Problems 21

3 Relativistic Mechanics   23
Introduction 23
3.1   Exploring Relativistic Momentum 23
3.2   Understanding Mass-Energy Equivalence 24
3.3   Understanding the Energy-Momentum Equation 25
3.4   Exploring Particle Decays 26
3.5   Understanding the Relativistic Doppler Effect 27
3.6   Exploring the Twin Paradox Using the Doppler Effect 28
Problems 29

II   The Need for a Quantum Theory   33

4   From Blackbody to Bohr   34
Introduction   34
4.1   Blackbody Radiation   34
4.2   Exploring Wien's Displacement Law   36
4.3   Brownian Motion   36
4.4   Exploring the e/m Experiment   37
4.5   Exploring the Millikan Oil Drop Experiment   38
4.6   Thomson Model of the Atom   40
4.7   Exploring Rutherford Scattering   42
4.8   The Bohr Atom and Atomic Spectra   43
4.9   Exploring the Emission Spectrum of Atomic Hydrogen   45
Problems   45

5   Wave-Particle Duality   48
Introduction   48
5.1   Wave and Particle Properties   48
5.2   Light as a Particle: Photoelectric and Compton Effects   50
5.3   Exploring the Properties of Waves   51
5.4   Wave Diffraction and Interference   52
5.5   The Electron Double-Slit Experiment   54
5.6   Double-Slit Experiment and Wave-Particle Duality   55
5.7   Exploring the Davisson-Germer Experiment   56
5.8   Diffraction Grating and Uncertainty   57
5.9   Phase and Group Velocity   58
5.10 Exploring the Uncertainty Principle   60
5.11 Exploring the Dispersion of Classical Waves   61
Problems   62

III   Quantum Theory   65

6   Classical and Quantum-mechanical Probability   66
Introduction   66
6.1   Probability Distributions and Statistics   66
6.2   Classical Probability Distributions For Moving Particles   68
6.3   Exploring Classical Probability Distributions   70
6.4   Probability and Wave Functions   70
6.5   Exploring Wave Functions and Probability   72
6.6   Wave Functions and Expectation Values   73
Problems   74

7   The Schrödinger Equation   76
Introduction   76
7.1   Classical Energy Diagrams   76

7.2   Wave Function Shape for Piecewise-constant Potentials 77
7.3   Wave Function Shape for Spatially-varying Potentials 79
7.4   Exploring Energy Eigenstates Using the Shooting Method 81
7.5   Exploring Energy Eigenstates and Potential Energy 82
7.6   Time Evolution 82
7.7   Exploring Complex Functions 84
7.8   Exploring Eigenvalue Equations 85
       Problems 86

8   The Free Particle   87
    Introduction 87
    8.1   Classical Free Particles and Wave Packets 87
    8.2   The Quantum-mechanical Free-particle Solution 88
    8.3   Exploring the Addition of Complex Waves 90
    8.4   Exploring the Construction of a Packet 91
    8.5   Towards a Wave Packet Solution 92
    8.6   The Quantum-mechanical Wave Packet Solution 93
    8.7   Exploring Fourier Transforms by Matching 95
    8.8   Exploring Wave Packets with Classical Analogies 96
           Problems 97

9   Scattering in One Dimension   99
    Introduction 99
    9.1   The Scattering of Classical Electromagnetic Waves 99
    9.2   Exploring Classical and Quantum Scattering 100
    9.3   The Probability Current Density 101
    9.4   Plane Wave Scattering: Potential Energy Steps 102
    9.5   Exploring the Addition of Two Plane Waves 105
    9.6   Plane Wave Scattering: Finite Barriers and Wells 106
    9.7   Exploring Scattering and Barrier Height 108
    9.8   Exploring Scattering and Barrier Width 109
    9.9   Exploring Wave Packet Scattering 110
           Problems 111

10 The Infinite Square Well   113
    Introduction 113
    10.1   Classical Particles and Wave Packets in an Infinite Well 113
    10.2   The Quantum-mechanical Infinite Square Well 114
    10.3   Exploring Changing Well Width 117
    10.4   Time Evolution 117
    10.5   Classical and Quantum-Mechanical Probabilities 119
    10.6   Two-State Superpositions 120
    10.7   Wave Packet Dynamics 122
    10.8   Exploring Wave Packet Revivals with Classical Analogies 124
           Problems 125

11 Finite Square Wells and Other Piecewise-constant Wells   127

Introduction 127

11.1 Finite Potential Energy Wells: Qualitative 127

11.2 Finite Potential Energy Wells: Quantitative 128

11.3 Exploring the Finite Well by Changing Width 132

11.4 Exploring Two Finite Wells 132

11.5 Finite and Periodic Lattices 133

11.6 Exploring Finite Lattices by Adding Defects 135

11.7 Exploring Periodic Potentials by Changing Well Separation 135

11.8 Asymmetric Infinite and Finite Square Wells 136

11.9 Exploring Asymmetric Infinite Square Wells 138

11.10 Exploring Wells with an Added Symmetric Potential 138

11.11 Exploring Many Steps in Infinite and Finite Wells 139

Problems 140

12 Harmonic Oscillators and Other Spatially-varying Wells   142

Introduction 142

12.1 The Classical Harmonic Oscillator 142

12.2 The Quantum-mechanical Harmonic Oscillator 143

12.3 Classical and Quantum-Mechanical Probabilities 146

12.4 Wave Packet Dynamics 147

12.5 Ramped Infinite and Finite Wells 148

12.6 Exploring Other Spatially-varying Wells 149

Problems 150

13 Multi-dimensional Wells   153

Introduction 153

13.1 The Two-dimensional Infinite Square Well 153

13.2 Two Particles in a One-dimensional Infinite Well 155

13.3 Exploring Superpositions in the Two-dimensional Infinite Well . . . 156

13.4 Exploring the Two-dimensional Harmonic Oscillator 157

13.5 Particle on a Ring 158

13.6 Angular Solutions of the Schrödinger Equation 161

13.7 The Coulomb Potential for the Idealized Hydrogen Atom 164

13.8 Radial Representations of the Coulomb Solutions 167

13.9 Exploring Solutions to the Coulomb Problem 168

Problems 170

IV   Applications   173

14 Atomic, Molecular, and Nuclear Physics   174

Introduction 174

14.1 Radial Wave Functions For Hydrogenic Atoms 174

14.2 Exploring Atomic Spectra 175

14.3 The H_2^+ Ion 177

14.4 Molecular Models and Molecular Spectra 178
14.5 Simple Nuclear Models: Finite and Woods-Saxon Wells 179
14.6 Exploring Molecular and Nuclear Wave Packets 180
Problems 181

15 Statistical Mechanics   **183**
Introduction 183
15.1 Exploring Functions: $g(\varepsilon)$, $f(\varepsilon)$, and $n(\varepsilon)$ 183
15.2 Entropy and Probability 184
15.3 Understanding Probability Distributions 186
15.4 Exploring Classical, Bose-Einstein, and Fermi-Dirac Statistics 188
15.5 Statistics of an Ideal Gas, a Blackbody, and a Free Electron Gas . . 189
15.6 Exploring the Equipartition of Energy 192
15.7 Specific Heat of Solids 194
Problems 195

Bibliography   **199**

Preface

By now it is hard to imagine an instructor who has not heard the call to "teach with technology," as it has resounded through educational institutions and government agencies alike over the past several years. It is, however, easier to imagine an instructor of modern physics and quantum mechanics who has not heard of the current research into the teaching and learning by Styer [1, 2], Robinett [3, 4], and others [5, 6, 7, 8, 9]. Despite this work, which focuses on improving the conceptual understanding of students, the teaching of quantum mechanics has remained relatively unchanged since its inception. Students, therefore, often see quantum physics in terms of misleading (such as the *convention* of drawing wave functions on potential energy diagrams) or incomplete visualizations, and as one dimensional and time independent (because of the focus on energy eigenstates in one dimension), and devoid of almost any connection with classical physics. These depictions short change quantum physics. Quantum physics is a far richer topic when non-trivial time evolution, multiple dimensions, classical-quantum connections, and research-based topics are discussed.

Physlet Quantum Physics is an interactive text with over 200 ready-to-run interactive exercises which use over 250 carefully-designed computer simulations for the teaching of quantum physics.[1] This material uses a standard easy-to-understand interface designed with a sound use of pedagogy in mind. The aim of *Physlet Quantum Physics* is to provide a resource for the teaching of quantum physics that enhances student learning through interactive engagement and visualization. At the same time, *Physlet Quantum Physics* is a resource flexible enough to be adapted to a variety of pedagogical strategies and local environments, covers a wide variety of topics, and is informed by current educational, experimental, and theoretical research.

CONTENT

Physlet Quantum Physics contains a collection of exercises spanning many concepts from modern and quantum physics. These exercises are based on computer animations generated in Java applets to show physics content. Every chapter of *Physlet Quantum Physics* contains three quite different Physlet-based exercises: Illustrations, Explorations, and Problems.

Illustrations are designed to demonstrate physical concepts. Students need to interact with the Physlet, but the answers to the questions posed in the narrative are given or are easily determined from interacting with it. Many Illustrations provide examples of quantum-mechanical applications. Other Illustrations are designed to introduce a particular concept or analytical tool. Typical uses of Illustrations would include reading assignments prior to class and classroom demonstrations. Illustrations are referred to in the text by their section number. For example, Section 8.6 covers free-particle quantum-mechanical wave packets.

---

[1] Previous simulation packages for quantum physics include Refs. [10, 11, 12, 13].

Explorations are tutorial in nature. They provide some hints or suggest problem-solving strategies to students in working problems or understanding concepts. Some narratives ask students to make a prediction and then check their predictions, explaining any differences between predictions and observations. Other Explorations ask students to change parameters and observe the effect, asking students to develop, for themselves, certain physics relationships (equations). Explorations appear in between the Illustrations in the text, making them an ideal test of knowledge gained from an Illustration or as a bridge exercise between two related concepts. Explorations can be used in group problem solving and homework or pre-laboratory assignments and are often useful as Just-in-Time Teaching exercises. Explorations are referred to in the text by their section number and their title begins with "Exploring." For example, Section 8.7 allows students to "explore" Fourier transforms by matching.

Problems are interactive versions of the kind of exercises typically assigned for homework. They require the students to demonstrate their understanding without as much guidance as is given in the Explorations. Some Problems ask conceptual questions, while others require detailed calculations. Typical uses for the Problems would be for homework assignments, in-class concept questions, and group problem-solving sessions. Problems appear at the end of each chapter.

PHYSLET CONDITIONS OF USE

Instructors may not post the Physlet-based exercises from *Physlet Quantum Physics* on the Web without express written permission from the Publisher and Mario Belloni, Wolfgang Christian, and Anne J. Cox for the English language, and in all other languages from Mario Belloni, Wolfgang Christian, and Anne J. Cox.

As stated on the Physlets Web site, Physlets (that is, the applets themselves) are free for noncommercial use. Instructors are encouraged to author and post their own Physlet-based exercises. In doing so, the text and script of Physlets-based exercises must be placed in the public domain for noncommercial use. Please share your work!

Authors who have written Physlet exercises and posted them on the Internet are encouraged to send us a short e-mail with a link to their exercises. Links will be posted on the Physlets page:

http://webphysics.davidson.edu/applets/Applets.html

More details can be found on the Conditions of Use page on the CD.

WEB RESOURCES

In addition to the interactive curricular material in this book and CD, instructors may also wish to view the *Physlet Quantum Physics Instructor's Guide*. The *Physlet Quantum Physics Instructor's Guide* is available for download is available for download from Prentice Hall's online catalogue pages at:

http://www.prenhall.com

Look for *Physlet Quantum Physics*.

BEFORE YOU START

Assigning *Physlet Quantum Physics* material without properly preparing the class can lead to frustration as small technical problems are bound to occur without testing. We use Physlets extensively in our quantum physics courses (modern physics and quantum mechanics) at Davidson College, but we always start the semester with a short tutorial whose sole purpose is to solve a problem in the way a physicist solves a problem; that is, to consider the problem conceptually, to decide what method is required and what data to collect, and finally to solve the problem. As a follow-up, we then assign a simple Physlet-based exercise that must be completed in one of the College's public computer clusters. This minimal preparation allows us to identify potential problems before Physlet-based material is assigned on a regular basis.

In response to these possible difficulties, we have written **Chapter 1: Introduction to Physlets**. This chapter provides students and instructors with a guided tutorial through the basic functionality of Physlets. After completing the exercises in Chapter 1, students and instructors alike should be in a position to complete the exercises in the rest of the book.

Before you begin, or assign material to students, you should also read the section on **Browser Tests and System Requirements**.

ACKNOWLEDGMENTS

There are a great many people and institutions that have contributed to our efforts, and we take great pleasure in acknowledging their support and their interest.

Some of the exercises that appear in the book and CD were originally created as part of an Associated Colleges of the South Teaching with Technology Fellowship with Larry Cain. We also thank Larry for providing many insightful comments and suggestions. We thank our students at Davidson College for testing of Physlet-based material in the classroom and the laboratory. Mur Muchane and the Davidson ITS staff have provided excellent technical support. We would also like to thank the Davidson College Faculty Study and Research Committee and Dean Clark Ross for providing seed grants for the development of Physlet-based curricular material. We also thank Nancy Maydole and Beverly Winecoff for guiding us through the grant application process.

M.B. would like to thank Ed Deveney, Mike Donecheski, Andy Gavrin, Tim Gfroerer, Laura Gilbert, Kurt Haller, Ken Krane, Bruce Mason, Rick Robinett, and Gary White for many useful and stimulating discussions regarding the teaching of quantum mechanics with and without Physlets.

W.C. would like to thank the numerous students who have worked with him over the years developing programs for use in undergraduate physics education. Some of our best Physlets are the result of collaborative efforts with student coworkers. In particular, we would like to single out Adam Abele, Cabell Fisher, and Jim Nolen.

A.J.C. would like to thank her colleagues at Eckerd College: Harry Ellis, Eduardo Fernandez, and Steve Weppner for their support and willingness to test Physlet-based materials in their classes. She also thanks her colleague (and father)

Bill Junkin for invaluable discussions about teaching modern physics and quantum mechanics, suggestions on presenting concepts, and his review of much of this text.

The following authors have contributed curricular material (unless otherwise stated, the author of the Illustration, Exploration, or Problem narrative is also the script author): Morten Brydensholt, Andrew Duffy, Francisco Esquembre, Bill Junkin, Steve Mellema, and Chuck Niederriter.

The following authors have contributed Java applets: Dave Krider and Slavo Tuleja.

We would like to thank Ernest Berringer, Pui-Tak Leung, Bruce Mason, Joseph Rothberg, and Chandralekha Singh for reviewing the manuscript.

We express our thanks to Erik Fahlgren, Christian Botting, and their coworkers at Prentice Hall for supporting the development of *Physlet Quantum Physics* and for all of their hard work getting this book to press on an accelerated schedule.

We also wish to express our sincerest thanks to those who have encouraged us the most, our spouses and our children:

Nancy and Emmy
Barbara, Beth, Charlie, and Rudy
Troy, Jordan, and Maggie

Part of this work was supported by a Research Corporation Cottrell College Science Award (CC5470) and three Associated Colleges of the South Teaching with Technology Fellowships. Physlets and Open Source Physics applets are generously supported by the National Science Foundation under contracts DUE-9752365 and DUE-0126439.

Java Tests and Browser Requirements

JAVA TESTS

Physlet Quantum Physics provides physics teachers and their students with a collection of ready-to-run, interactive, computer-based curricular material for the teaching of modern physics and quantum mechanics. All that is required is the *Physlet Quantum Physics* CD, the latest version of Java from Sun Microsystems, and a browser that supports Java applets and JavaScript-to-Java communication (LiveConnect). This combination is available for recent versions of Microsoft Windows, for most versions of Unix (such as Linux), and for the latest version of the Macintosh operating system (see Macintosh instructions below). Although we occasionally check Physlets using other combinations, Microsoft Windows 2000 and XP with Internet Explorer (IE) or the Mozilla browsers are our reference platforms.

To check whether your computer already has Java installed, go to the Preface Chapter on the CD and navigate to the Java Tests and Browser Requirements page. There you will find three buttons. Click the "Check for Java" buttons to

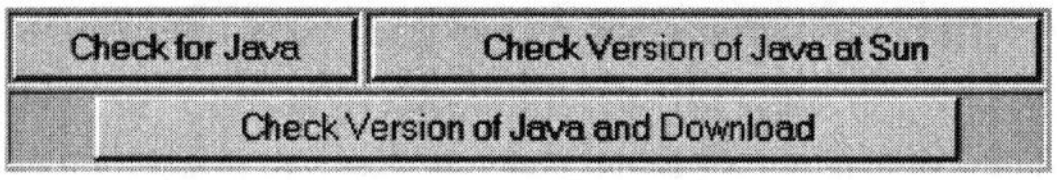

Check for Java	Check Version of Java at Sun
Check Version of Java and Download	

FIGURE 1: The "Check for Java," "Check the Version of Java at Sun," and "Check Version of Java and Download" buttons.

see if Java is on your machine, and if you have an internet connection, click the "Check the Version of Java at Sun" button. If your browser fails the Java test, or if your version of Java is older than 1.4.2, you may use the "Check Version of Java and Download" button to get the latest version of Java from Sun. If your browser passes the Java test, and is using Java version 1.4.2 or later (version 1.5.0 or later is preferred), it is ready to use all of the interactive material on the CD, and you may skip the following Sections.

GETTING JAVA

For Windows, Unix, and Linux operating systems, the Sun Java virtual machine (JVM or Java VM) is downloadable from the Java Web site:

$$\text{http://java.sun.com}$$

After downloading the file to your hard drive, double-click on its icon to run the installer. Follow the instructions the installer provides.

Although it is possible to simultaneously install Java VMs from Microsoft and Sun Microsystems on Windows computers, a browser can only run one VM at a time. You can switch between these two JVMs in Internet Explorer. Start Internet Explorer and click the Advanced tab under Tool—Internet Options from the Internet Explorer menu bar. The following dialog box shown in Figure 2 will appear.

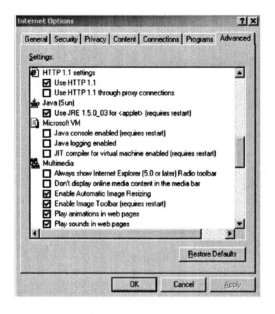

FIGURE 2: The advanced Internet Options dialog box accessed from within Internet Explorer.

Figure 2 shows that this computer has two Java VMs and that it is currently configured to run the Sun VM (JRE 1.5.0_03). The option for the Java (Sun) VM will not appear unless the Sun Java Runtime Environment has been installed. You will need to close all browser windows if you decide to switch VMs. You do not, however, need to restart the computer.

For the Macintosh platform, the Apple JVM (called MRJ, Macintosh Runtime Java) is available from Apple and is currently available in versions 1.4.2 and 1.5.

WINDOWS BROWSERS

While the latest version of Internet Explorer runs Java applets very well, Mozilla, and Mozilla Firefox browsers offer an open-source alternative on the Windows operating system. You can download the Mozilla and Firefox browsers from the Mozilla Web site:

`http://www.mozilla.org`

After downloading the file to your hard drive, double-click on its icon to run the installer. Follow the instructions the installer provides. The Mozilla and Mozilla Firefox browsers require that the Sun JVM be installed on your computer.

MACINTOSH BROWSERS

Apple currently supports LiveConnect using the latest Safari browser under the OSX Panther (10.3) and OSX Tiger (10.4) operating systems. Although both Java 1.1 and JavaScript are implemented in older Macintosh operating systems, the ability of JavaScript to communicate with a Java applet is problematic on these older Apple computers. **This problem is partially solved with Panther/Tiger and Safari.** At the time of this printing, Mozilla and Firefox do not support LiveConnect on the Macintosh.

We have found that while individual html pages on the CD run, browsing the contents of the CD within Safari on Macintosh OSX is problematic. A web page has properly loaded when the red message on the html page:

```
Please wait for the animation to completely load.
```

vanishes. If the message vanishes the page is ready to run. If not, you must quit Safari and navigate through the CD contents to the exercise you want to complete using the file browser (Finder). Only then should you open the html page in Safari. Once you have completed the exercises on a particular html page, you must quit Safari to open a subsequent page. This process allows limited functionality of the CD contents on Macintosh OSX.

LINUX BROWSERS

Mozilla, and Mozilla Firefox on Unix/Linux operating systems support LiveConnect. You can download Mozilla browsers from the Mozilla Web site:

```
http://www.mozilla.org
```

After downloading the file to your hard drive, double-click on its icon to run the installer. Follow the instructions the installer provides. The Mozilla and Mozilla Firefox browsers require that the Sun JVM be installed on your computer.

C H A P T E R 1

Introduction to Physlets

1.1 STATIC IMAGES VERSUS PHYSLET ANIMATIONS
1.2 ANIMATIONS, UNITS, AND MEASUREMENT
1.3 EXPLORING THE INPUT OF DATA: NUMBERS
1.4 EXPLORING THE INPUT OF DATA: FORMULAS
1.5 GETTING DATA OUT
1.6 EXPLORING THE INPUT OF DATA: COMPLEX EXPRESSIONS

INTRODUCTION

This chapter serves as an introduction to the various types of interactive curricular material you will find in *Physlet Quantum Physics*. In addition, this chapter gives a brief tutorial on the types of basic computer skills you will need to run, interact with, and complete the exercises.

1.1 STATIC IMAGES VERSUS PHYSLET ANIMATIONS

A Physlet is a **Phys**ics (Java) App**let** written at Davidson College. We use Physlets to animate physical phenomena and to ask questions regarding the phenomena. Sometimes you will need to collect data from the Physlet animation and perform calculations in order to answer the questions presented. Sometimes simply viewing the animation will be enough for you to complete the exercise.

The Physlet animations presented in *Physlet Quantum Physics* will be similar to the static images in your textbook. There are differences that need to be examined, however, because we will be making extensive use of these types of animations throughout *Physlet Quantum Physics*. First consider the following image.[1]

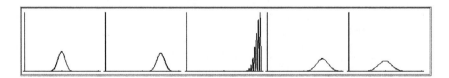

FIGURE 1.1: Images of the time evolution of the position-space probability density corresponding to a quantum-mechanical wave packet at times near its collision with an infinitely hard wall.

---

[1]One of the first such computer-generated images was discussed by A. Goldberg, H. M. Schey, and J. L. Schwartz, "Computer-generated Motion Pictures of One-dimensional Quantum-mechanical Transmission and Reflection Phenomena," *Am. J. Phys.* **35**, 177-186 (1967).

This image represents the time evolution of the position-space probability density corresponding to a quantum-mechanical wave packet as it approaches, collides with, and reflects from an infinitely hard wall. The packet has an initial momentum to the right and the images are shown at equal time intervals. We are supposed to imagine the motion of the packet by reading the images from left to right. We notice that as the packet moves to the right its leading edge encounters the infinite wall first and is reflected back towards the middle of the well. There are several interesting features about this problem that are not depicted in the image however. While we get the general sense of what happens, we lose a lot of the specific information. This is especially apparent at the times when the packet is in contact with the wall.[2]

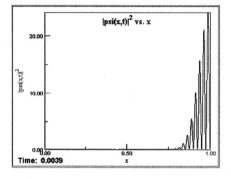

FIGURE 1.2: Image of the animation of the position-space probability density of a quantum-mechanical wave packet at a time just after its collision with an infinitely hard wall.

Now consider the Physlet animation of the same situation. Press the "play" button to begin the animation. Note that the VCR-type buttons beneath the animation control the animation much like buttons on a VCR, CD, or DVD player. Specifically:

- **play**: starts the animation and continues it until either the animation is over or is stopped.
- **pause**: pauses the animation. Press play to resume the animation.
- **step>>**: steps the animation forward in time by one time step.
- **<<step**: steps the animation backward in time by one time step (the size of the time step varies with the animation). In this animation there is no "<<step" button.
- **reset**: resets the animation time to the initial time. Then, press "play" to run the animation from the beginning.

Make sure you understand what these buttons do, since you will need to use

---

[2]Such a collision is often called a quantum bounce. See, for example, M. Andrews, "Wave Packets Bouncing Off of Walls," *Am. J. Phys.* **66** 252-254 (1998), M. A. Doncheski and R. W. Robinett, "Anatomy of a Quantum 'Bounce,'" *Eur. J. Phys.* **20**, 29-37 (1999), and M. Belloni, M. A. Doncheski, and R. W. Robinett, "Exact Results for 'Bouncing' Gaussian Wave Packets," *Phys. Scr.* **71**, 136-140 (2005).

them throughout the rest of the book when you interact with the Physlets on the CD.

In addition to these buttons, there are hyperlinks on the page that control which animation is played. For example, on the html page associated with this exercise, **Restart** reinitializes the applet to the way it was when the page was loaded. On other pages there will often be a choice of which animation to play, but **Restart** always restores the animation to its initial condition.

So why animation in addition to static images? Most of the examples typically studied in quantum physics are often limited to a few simple exactly solvable problems. Interactive animations allow for the study of a wide variety of different situations, many not solvable analytically. In addition, the time evolution and dynamics of quantum-mechanical systems are often difficult to understand if you are trying to describe it with a static picture (or even a series of static pictures). Because the examples in this book are interactive animations, you can actually see the details of the quantum-mechanical time development and change the initial conditions to explore other scenarios.

Restart (or "reset") the animation and play it again. Watch the collision with the right wall in detail. What do you notice about the motion of the packet? First, notice that the packet spreads with time. This is a quantum-mechanical effect (see Chapter 8 for details). Also note that the collision with the wall is rich in detail that may be missed if you only have one image of the collision.

1.2   ANIMATIONS, UNITS, AND MEASUREMENT

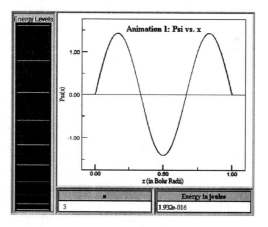

FIGURE 1.3: The $n = 5$ energy eigenstate for a particle in an infinite square well. Alongside the wave function is the energy spectrum.

Most physics problems are idealizations of actual physical situations. These idealizations can take many forms in quantum physics. One way we can simplify a problem is by the choice of convenient units. This is especially important in quantum physics and in the animations we show in *Physlet Quantum Physics*.

Consider **Animation 1** in which we have an electron ($m_e = 9.109 \times 10^{-31}$ kg) confined to a one-dimensional box with a length of the Bohr radius ($a_0 = 5.3 \times 10^{-11}$ m). According to quantum mechanics (see Chapter 10), the energy spectrum for such a situation is quantized such that

$$E_n = \frac{n^2 \pi^2 \hbar^2}{2mL^2} \, , \tag{1.1}$$

where $\hbar = 1.055 \times 10^{-34}$ J·s is Planck's constant divided by 2π, L is the length of the box (here one Bohr radius), and n is a positive integer. For the current situation, the energy spectrum is $E_n = n^2(2.146 \times 10^{-17})$ J, which is depicted in **Animation 1**. You can **click-drag** in the energy spectrum on the left of the animation to change the energy level (only the first 10 are shown) and as you do so, the displayed energy level turns from green to red. What do you notice about the energy spectrum? You should notice that the numerical value for the energy is not very helpful in determining the functional form of the energy levels if you did not know it already. Can we make the functional form of the energy spectrum more transparent?

We can use units that make the physics more transparent by scaling the problem accordingly. This entails setting certain variables in the problem to simpler values (such as 1) and working in a dimensionless representation. In **Animation 2** we have used one such common choice of units in numerical simulations: $\hbar = 2m = e^2 = 1$.[3] In these units, the length scale is $a_0/2$ (0.265×10^{-10} m) and the energy scale is 4 Rydbergs or 2 Hartrees (54.4 eV). Notice that the energy spectrum is now somewhat simpler: at the very least we do not have a factor of 10^{-17} in the energy anymore. We can also use other scaling conditions to simplify the energy spectrum even further as shown in **Animation 3**. Here we set the combination $\pi^2 \hbar^2 / 2mL^2 = 1$ which now scales the energy spectrum in units of the ground-state energy. What can you say about the energy spectrum now? By choosing appropriate units, it has become clear that the energies are an integer squared times the ground-state energy.

In general, you should look for the units specified in the problem (whether from your text or from *Physlet Quantum Physics*): **On the *Physlet Quantum Physics* CD all units are given in boldface in the statement of the problem.**

Although computer simulations allow precise control of parameters, their spatial resolution is not infinite. Whenever these data are presented on screen as numeric values, they are correct to within the last digit shown. Start **Animation 4** by clicking the "set the state: t=0" button and follow the procedures below to make position measurements. What is shown in the animation is the time development of the same states shown in the previous animations ($\hbar = 2m = 1$).

Start by **click-dragging** the mouse inside the animation to make measurements. Try it. Place the cursor in the wave function graph and hold down the left mouse button. Now drag the mouse around to see the x and y coordinates of the

---

[3]It is also common to see the convention that $\hbar = m = e^2 = 1$, which are called *atomic units* and is due to Hartree. In atomic units, the distance scale is given in Bohr radii ($a_0 = \hbar^2/me^2 = 1$) and energy is given in Hartrees (1 Hartree = 2 Rydbergs = $me^4/\hbar^2 = 1$). We choose $\hbar = 2m = 1$ because the combination $\hbar^2/2m$ occurs in the Schrödinger equation.

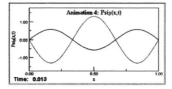

FIGURE 1.4: The real and imaginary components of the time-dependent ground-state wave function for a particle in an infinite square well.

mouse change in the lower left-hand corner of the animation. Notice the way the coordinates change. In addition, **these measurements cannot be more accurate than one screen pixel**. This means that depending on how you measure the position of an object you may get a slightly different answer than another student in your class.

In this animation we have also given you two choices for the time scale. In the default animation, we have not changed the time scale and the ground state takes $2/\pi = 0.6366$ to go back to its $t = 0$ position. When you click in the check box and click the "set the state: t=0" button, you get the same animation, but with the energy, and therefore the time, scaled. We often do this to make time measurements easier to understand. In this example we have scaled the time so that the ground state returns to its $t = 0$ position when $t = 1$. This time is the so-called *revival time* for the infinite square well.[4]

1.3   EXPLORING THE INPUT OF DATA: NUMBERS

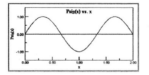

FIGURE 1.5: An energy eigenstate for a particle in an infinite square well.

Shown is the time-independent wave function for a particle in an infinite square well of length $L = 2$ ($\hbar = 2m = 1$). You may change the state by choosing an n. **Restart**.

Click the "set value" button. Now change the value of the quantum number, n, by typing in the text box and then click the "set value" button again.

(a) Find the limits of the values you can type in the text box.

(b) Have you tried non-integer values? If not, try a few.

(c) Why do you think these values have been chosen?

(d) Now try typing in "abcd". What happens?

---

[4]This is discussed in Chapter 10.

1.4   EXPLORING THE INPUT OF DATA: FORMULAS

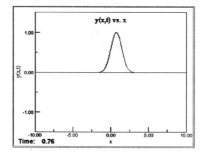

FIGURE 1.6: A time-dependent Gaussian wave form.

In many animations you will be expected to enter a formula to control the animation **(position is given in centimeters and time is given in seconds)**. **Restart**. In this Exploration, you are to enter a classical wave function, $y(x, t)$, for a wave on a string into the text box. This initializes the wave function at $t = 0$ on the graph. Once you have done this, the time evolution of the wave function is governed by the form of the wave function and the "resume" and "pause" buttons.

There are a few important rules for entering functions. Notice that the default value in the text box, `exp(-(x-t)*(x-t))`, corresponds to the wave function $e^{-(x-t)(x-t)}$. Note that the product in the argument of the exponential is `(x-t)*(x-t)` and **NOT** `(x-t)(x-t)`. This is the way the computer understands multiplication. You must enter the multiplication sign, `*`, every time you mean to multiply two things together. Remove the `*` and see what happens. You get an error and you can see what you entered. Division is represented as `x/2` and **NOT** `x\2`. In addition, the Physlet understands the following functions:

```
sin(a)  cos(a)   tan(a)   sinh(a)   cosh(a)  tanh(a)
asin(a) acos(a)  atan(a)  asinh (a) acosh(a) atanh(a)
step(a) sqrt(a)  sqr(a)   exp(a)    ln(a)    log(a)
abs(a)  ceil(a)  floor(a) round(a)  sign(a)  int(a)    frac(a)
```

where "a" represents the variable expected in the function (here it is x, t, or a function of the two).

Try the following real functions for $y(x, t)$ and describe what each change does to the time-dependent wave function:

(a) `exp(-(x-2*t)*(x-2*t))`

(b) `exp(-(x+2*t)*(x+2*t))`

(c) `exp(-(x-4*t)*(x-4*t))*cos(2*x)`

(d) `sin(2*pi*x-2*pi*t)`

(e) `sin(pi*x)*cos(pi*t)`

Try some other functions for practice.

1.5   GETTING DATA OUT

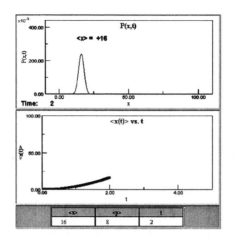

FIGURE 1.7: An accelerating quantum-mechanical Gaussian wave packet with a graph of its position expectation value, $\langle \hat{x} \rangle$, and a table with the position and momentum expectation values of the packet with the current time.

In Section 1.2 you learned about units and how to click-drag in an animation to get data from the animation. Here we will discuss several other ways in which data are depicted in animations.

Press the "play" button to begin. Shown in the animation ($\hbar = 2m = 1$) is the probability density corresponding to an accelerating Gaussian wave packet. When you press the "play" button, the packet will move across the screen in a predefined way (obeying the rules of quantum mechanics). Along with the probability density are depictions of the packet's average position: as an on-screen numerical statement, as data in a table, and as a function of time on a graph. You may of course click-drag in the animation to measure average position (since a Gaussian is symmetric about its maximum value) and amplitude as well.

Why do we show all of these different representations? Because they provide complementary ways of thinking about the phenomena. Click **Restart** and play the animation again. Notice how the different representations of the motion change with the motion of the packet. With a lot of practice, physicists can look at the motion of an object and can tell you the various properties of the motion. How do we do that? By having different mental representations in our heads. Specifically,

- **on-screen numerical statements of position** facilitate the measurement process as the value is always given. These statements can be for any variable, not just position.
- **data tables** are used to compare two or more values that are changing like in the animation where $\langle \hat{x} \rangle$, $\langle \hat{p} \rangle$, and t are changing.
- **graphs** are used to summarize all of the data corresponding to the motion of an object that occurs during a time interval. The graph summarizes all of the data shown in the on-screen calculation and the data table. When you

get a good-looking graph, you can usually right-click on it to clone the graph and resize it for a better view. Try it!

In practice, we will never set up an animation to give you all of these depictions simultaneously. We usually pick one or two representations that best represent the phenomena.

Note that some animations depict motion that started before the animation begins and continues beyond the time that the animation ends. In the animations on this page, the packet starts at rest at $t = 0$ but continues its motion beyond the $t = 5$ mark when the animation ends.

1.6   EXPLORING THE INPUT OF DATA: COMPLEX EXPRESSIONS

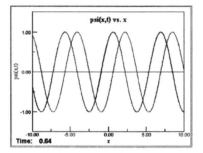

FIGURE 1.8: A quantum-mechanical plane wave moving to the right showing its real and imaginary components.

Complex functions are absolutely necessary to describe quantum-mechanical phenomena. This is due to quantum-mechanical time evolution being governed by the Schrödinger equation[5] which guarantees complex solutions. In many exercises you will be expected to enter a formula to control the animation **(position and time are given in arbitrary units)**. <u>Restart</u>. In this Exploration, you are to enter the real (the blue curve on the graph) and imaginary (the pink curve on the graph) parts of a function, $f_{\text{Re}}(x, t)$ and $f_{\text{Im}}(x, t)$, for $t = 0$. Once you have done this, the time evolution of the function is governed by the form of the function you have chosen and the "resume" and "pause" buttons.

Besides entering $[f_{\text{Re}}(x, t), f_{\text{Im}}(x, t)]$, the real and imaginary components of the function, you will also be asked to enter the function in its magnitude and phase form, $f(x, t) = A(x, t) e^{i\theta(x,t)}$ where A and θ are real functions. The default function for this Exploration is $[\cos(x - t), \sin(x - t)]$ or $f(x, t) = e^{i(x-t)}$ which is called a plane wave.[6] In the text box you can enter a complex function in magnitude and phase form. Try it for the plane wave, `exp(i*(x-t))`, to see if you get the same picture as above.

---

[5]By the Schrödinger equation we mean what is often called the time-dependent Schrödinger equation since this is *the* Schrödinger equation.

[6]For example, the complex function $z(x) = e^{ix} = \cos(x) + i\sin(x)$ and $z(x) = 1/(x + i) = x/(x^2 + 1) - i/(x^2 + 1)$.

Input the following functions for the real and imaginary parts of $f(x, t)$ in the first animation, then determine what amplitude and phase form you have to enter into the text box of the second animation to mimic the results you saw in the first animation.

(a) `real = exp(-0.5*(x+5)*(x+5))*cos(pi*x)`
`imaginary = exp(-0.5*(x+5)*(x+5))*sin(pi*x)`

(b) `real = sin(2*pi*x)*cos(4*t)`
`imaginary = sin(2*pi*x)*sin(4*t)`

Try some other complex functions for practice.

PROBLEMS

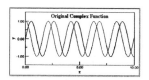

1.1. Shown in the top animation is a complex time-dependent quantum-mechanical wave function with the real part shown in blue and the imaginary part shown in pink. Determine the wave function (real and imaginary) of the original wave function. Enter the formula for this wave function in the text boxes in order to match it with the top wave function. Once you have done so, press the "import function and play" button to see if you determined the correct form of the original wave function. **Restart**.

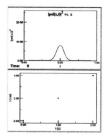

1.2. A quantum-mechanical wave packet moves in an infinite well with walls at $x = 0$ and $x = 1$ as shown in the animation **(position and time are given in arbitrary units)**. Take data from the animation and create a position vs. time graph of the central peak of the Gaussian wave function. To add data to the graph, type your (t, x) data into the text boxes and then click the "add datum" button. Use the "clear graph" button to start another graph. **Restart**.
(a) Describe in words the shape of the curve that will fit through your data.
(b) How accurate were your measurements when the packet was near the wall?
(c) What does this function tell you about the short-term behavior of the wave packet?
(d) How does it compare to what you might expect to see for the position of a classical particle in a similar infinite well?

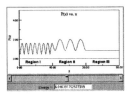

1.3. Shown is the probability density corresponding to a quantum-mechanical plane wave traveling to the right. The plane wave experiences constant potential energy step (a finite barrier) between $x = 10$ and $x = 20$. The probability density of a pure plane wave is always a constant (flat). Therefore, when the probability density is a constant to the left of the potential energy barrier, the wave is 100% transmitted. When there is reflection, the reflected plane wave adds to the incident plane wave, and a standing wave pattern can develop when the amplitude of the reflected wave is the same as the incident wave **Restart**.

  (a) Drag the slider to change the plane wave's energy. With what energy do you first notice a transmitted wave? Note: in this situation, transmission is nonzero for any value of the initial plane wave's energy.

  (b) Determine the energies that give you 100% transmission. Note: you will not be able to get the energy of the first resonance exactly since this resonance is too sharp compared to the slider's resolution.

  (c) For the energies that give you 100% transmission, what do you notice about the pattern of the probability density between $x = 10$ to $x = 20$?

One-dimensional scattering is covered in detail in Chapter 9.

PART ONE

SPECIAL RELATIVITY

C H A P T E R   2

Space and Time in Special Relativity

2.1   SYNCHRONIZING CLOCKS
2.2   EXPLORING SYNCHRONIZING CLOCKS BY VIEWING
2.3   SIMULTANEITY
2.4   LIGHT CLOCKS, TIME DILATION, AND LENGTH CONTRACTION
2.5   UNDERSTANDING SPACETIME DIAGRAMS
2.6   EXPLORING WORLDLINES BY MATCHING
2.7   EXPLORING THE POLE AND BARN PARADOX
2.8   EXPLORING THE TWIN PARADOX

INTRODUCTION

Special relativity, along with quantum mechanics, can be considered a cornerstone of modern physics. In fact, the combination of special relativity and quantum mechanics yields relativistic quantum mechanics. While this book is not a book on relativistic quantum mechanics,[1] the knowledge of basic relativity is important for the understanding of quantum mechanics.

2.1   SYNCHRONIZING CLOCKS

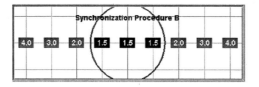

FIGURE 2.1: A configuration of clocks created to depict a reference frame in special relativity. Each clock has a built in time delay and only turns on when a light pulse from the master clock reaches it.

One of the most important ideas regarding how we measure the properties of moving objects is the simultaneity of events (or lack thereof) and the synchronization of clocks. However, no description of special relativity can begin without the introduction of a reference frame in which to perform measurements. We construct a reference frame with numerous clocks placed at 1-meter increments (position is

---

[1]See for example, J. J. Sakurai, *Advanced Quantum Mechanics* Addison-Wesley (1967).

given in meters and time is given in in the time it takes light to travel one meter or 3.33×10^{-9} seconds). If we did this in three dimensions we would have a cubic lattice spanning all space. Let's simplify matters by only considering one dimension. We want all of the clocks to be synchronized with a master clock at the origin.

One way to do this is to synchronize all of the clocks at the master clock and then slowly move each clock into place on the spacetime lattice. This is the procedure depicted in "Synchronization Procedure A." Notice that it takes some time for the animation to complete, as we transport the clocks slowly so as to not incur any significant time-dilation errors.

A second procedure is depicted in "Synchronization Procedure B." We know that for every one meter a clock is displaced from the master clock, there is going to be one meter of light travel time (or 3.33×10^{-9} seconds) delay. We build this time delay into all of our clocks, put them in place, and start them when a light pulse from the master clock reaches them.

With all of the clocks now synchronized, we can start analyzing events in this frame of reference. We first note several properties of this and every other reference frame.

- It is of infinite extent. While we have drawn the reference frame in only one dimension and of a finite size, it is actually three dimensional and infinite.

- All observers (so-called intelligent observers[2]) in this reference frame agree on the simultaneity of events. These observers take into account the light travel time and it is this time that is recording in their laboratory notebooks. Since we are able to look at all of space at once in the animation, we can be considered omnipresent observers. See Section 2.3 for more details on simultaneity.

2.2   EXPLORING SYNCHRONIZING CLOCKS BY VIEWING

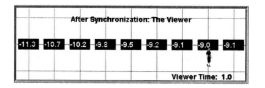

FIGURE 2.2: A synchronized spacetime lattice (a reference frame) as seen by a viewer of the clocks who does not factor in light travel time delay.

What we, as omnipresent observers, are recording in our laboratory notebooks after the synchronization of clocks is not what a viewer who sees the lattice of clocks from a fixed point in space would see. A viewer would not see all of the clocks synchronized because of the light-travel time delay. Instead he/she would

---

[2]See for example, R. E. Scherr, P. S. Shaffer, and S. Vokos, "The Challenge of Changing Deeply Held Student Beliefs about the Relativity of Simultaneity," *Am. J. Phys.* **70**, 1238 (2002) and R. E. Scherr, P. S. Shaffer, and S. Vokos, "Student Understanding of Time in Special Relativity: Simultaneity and Reference Frames," *Phys. Educ. Res., Am. J. Phys.* Suppl. **69**, S24 (2001).

see what is depicted by a "viewer." Note that when we talk about reference frames and synchronized clocks, we do so in the sense of what an intelligent observer or an omnipresent observer would observe (the light travel-time delay is removed). Often students (and physicists alike) mistakenly believe that all of the strange things that are a part of special relativity are due to light-travel time delay; they are not.

Vary the x and z positions of the viewer to see the effect of light travel time on the clock readings as seen by the viewer (position is given in meters and time is given in the time it takes light to travel one meter or 3.33×10^{-9} seconds).

(a) How do the clock readings change as you vary the x position of the viewer?

(b) How do the clock readings change as you vary the z position of the viewer?

(c) Where does a viewer need to be located in order for the clocks to appear (approximately) synchronized?

2.3 SIMULTANEITY

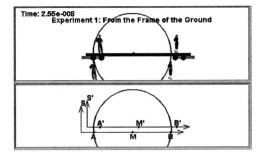

FIGURE 2.3: An experiment showing a light pulse emitted from the center of a railcar as seen from the reference frame of the ground. The pulse reaches both ground observers at the same time but does not reach the railcar observers at the same time as seen from the frame of the ground.

In these animations, a lightning bolt hits the center of *two different* flatbed railcars at $t = 0$ seconds (position is given in meters and time is given in in the time it takes light to travel one meter or 3.33×10^{-9} seconds). There is a relative velocity between the reference frame of the ground, called S in the lower panel of the animation, and the reference frame of the railcar, called S' in the lower panel of the animation. In the animation "Experiment 1: From the Frame of the Ground," the event of the lightning strike and the subsequent transmission of this information are shown as observed from the reference frame of the ground. Alternatively, in the animation "Experiment 2: From the Frame of the Railcar," a second, and different, experiment is depicted as observed from the reference frame of a second railcar.

For ease of viewing, there is an offset in the y direction shown between S and S' in the animations. Likewise there is an offset between the people on the railcars and the observers on the ground.

Begin by reconsidering what is observed in the "From the Frame of the Ground" by playing and pausing the animation. In this frame of reference, the railcar is moving to the right at a given speed as depicted in the lower panel by S' moving to the right. As the circle representing the path of a spherical light wave expands, it first encounters the observer (the man) at A'. This means according to any intelligent observer in reference frame S, no matter where he or she is in the reference frame, this event (the light reaching A') happens first. Next the outgoing spherical light wave reaches the intelligent observers at A and B simultaneously (again as observed by any intelligent observer in reference frame S no matter where he or she is in the reference frame). Finally, the light reaches the observer at B'.

Now consider what is observed in the "From the Frame of the Railcar" by again playing and pausing the animation. In this frame of reference (S') the railcar is stationary and the ground is moving to the left at a given speed as depicted in the lower panel by S moving to the left. As the circle representing the path of a spherical light wave expands it first encounters the observer (the woman) at B. This means according to any intelligent observer in reference frame S', no matter where he or she is in the reference frame, this event (the light reaching B) happens first. Next the outgoing spherical light wave reaches the intelligent observers at A' and B' simultaneously (again as seen by any intelligent observer in reference frame S' no matter where he or she is in the reference frame). Finally, the light reaches the observer at A.

When we are dealing with moving reference frames, we must modify our idea of simultaneity to include the idea that events that are simultaneous in one reference frame are not simultaneous in another reference frame. This is perhaps one of the most important things to keep in mind when considering the apparent paradoxes that arise in special relativity. Almost all of these apparent paradoxes can be understood by remembering that events simultaneous in one reference frame are not simultaneous in another reference frame.

Also note that there are two different experiments depicted. To see why this must be the case, we first need to discuss length contraction, which follows in Section 2.4.

2.4 LIGHT CLOCKS, TIME DILATION, AND LENGTH CONTRACTION

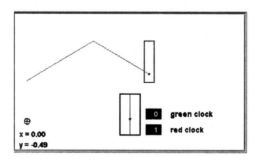

FIGURE 2.4: Two light clocks, one stationary and one moving. The light pulses bounce off of mirrors in the top and bottom of each clock.

One of the best ways to visualize time dilation and length contraction of moving objects is with the construction of a light clock. A light clock consists of a box with a light pulse emitter and a light detector at its bottom wall and a mirror on its top wall. Light is emitted from the bottom wall and every time the light pulse returns to the bottom wall the detector triggers a tick of the clock and another light pulse is emitted (position is given in meters by dragging the cross-haired cursor around the animation). The total vertical distance traveled by the light for the stationary clock is L_0, where $L_0/2$ is the distance between the walls of the clock. For simplicity, in this animation the distance between the walls is 0.5 meters and therefore the total vertical distance is L_0 or 1.0 meter. Given this, the clock reads time in meters. What does this mean? Light travels one meter every 3.33×10^{-9} seconds. Therefore every click measures the time it takes for light to travel 1 meter or 3.33×10^{-9} seconds.

Now consider what a stationary observer relative to a moving (green) light clock records. Set β ($= v/c$) to 0.5 and press the "set value and play" button. The green clock moves at half the speed of light (ignore the length contraction of the horizontal size of the light clock as it is irrelevant for this discussion). Given Einstein's postulate about the constancy of the speed of light, the moving (green) clock (as recorded by the stationary observer) must tick slower than that of the stationary (red) clock.

This result occurs for the light clock because the speed of light is constant in any reference frame; therefore the distances traveled by the two light pulses must be the same as viewed in the frame of the stationary clock. However, the distance traveled by the moving clock, as seen from the point-of-view of an observer in the stationary clock's reference frame, involves both horizontal and vertical components, and it is only the vertical component of the light pulse's motion that contributes to the clock ticks (as seen from the reference frame of the clock at rest in the animation). We can calculate these distances by using the Pythagorean theorem:

$$(c\Delta t')^2 = (v\Delta t')^2 + L_0^2 , \qquad (2.1)$$

where $\Delta t'$ is the time interval that an observer in the stationary frame sees the light travel time to be. We can simplify this equation to $(\Delta t')^2 = (\beta \Delta t')^2 + (L_0/c)^2$ by dividing by the speed of light. By grouping common terms we find that:

$$(1 - \beta^2)\Delta t'^2 = (L_0/c)^2 = \Delta t^2 , \qquad (2.2)$$

since $\Delta t = L_0/c$ for the stationary clock (and for the moving clock as observed in the moving clock's frame of reference). Therefore $\Delta t' = \gamma \Delta t$ where

$$\gamma = 1/\sqrt{1 - \beta^2} . \qquad (2.3)$$

Therefore, it takes more clicks as measured by the stationary clock to measure a time interval of a moving clock. Observed from stationary frames, moving clocks run slower. This is called time dilation.

Note that we are talking about what is recorded by an observer in the stationary frame, and not what the moving observer records. The time interval of a stationary clock remains Δt (whether it is the red clock in the stationary frame or the green clock as recorded in its reference frame).

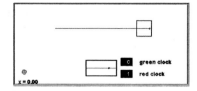

FIGURE 2.5: Two rotated light clocks, one stationary and one moving. The light pulses bounce off of mirrors in the left and right walls of each clock.

In this animation, the light clocks are rotated 90 degrees (position is given in meters by dragging the cross-haired cursor around the animation). The results we saw for the time dilation still occurs for the rotated clocks. Therefore, given the fact that in the stationary frame the moving clock still ticks slower, the only way for this to happen is for the distance between the walls of the moving clock to be contracted. In fact, the clock must appear to be contracted to $L' = L_0/\gamma$, where L_0 is the length of the clock as seen in its own (stationary) reference frame and L' is the length of the moving clock as seen from the stationary frame.[3]

2.5   UNDERSTANDING SPACETIME DIAGRAMS

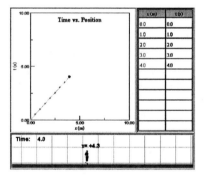

FIGURE 2.6: A postion vs. time graph as the first step toward a spacetime diagram.

One of the most useful ways to visualize moving objects in special relativity is with spacetime diagrams. Consider the animation of a woman walking (position given in meters and time is given in seconds). The motion of the woman in "Animation 1" is rather ordinary and is plotted on a position versus time graph as well as in a data table. Notice that the speed of the woman can be determined by the slope of the position versus time graph (in this case 1 m/s).

Now consider what is being represented in "Animation 2." In this animation, time is plotted versus position. The graph is the same as Animation 1 with the axes flipped. This way to represent the motion of the woman is *almost* what physicists

---

[3]For all of the algebra, see pages 26-29 of K. Krane, *Modern Physics* 2nd ed., John Wiley and Sons, New York (1996).

would call a spacetime diagram. Two things are missing: we want to treat the time on the same footing (as far as units) as the position, and we need to take into account the universal speed limit, c.

"Animation 3" puts time on equal footing with position by multiplying time by the speed of the woman. Therefore, velocity of the woman times time is plotted versus position. This converts the unit of time into meters.

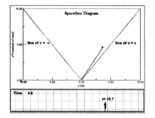

FIGURE 2.7: A spacetime diagram for a woman traveling a constant velocity near the speed of light.

Finally, we need to take into account the universal speed limit of the speed of light. For a true spacetime diagram, we multiply the time by the speed of light. The unit of the y axis now becomes the amount of time it takes for light to travel one meter or 3.33×10^{-9} seconds. Select β to be zero and then press the set value and play button. Notice that the woman does not move in space but moves in time. Now try a β of 0.9. What does her *trajectory* or *worldline* on the spacetime diagram look like now? As $|\beta|$ gets bigger (approaches 1) the trajectory of the woman on the spacetime diagram approaches the line of $v = c$. This is the 45-degree line of slope 1 that appears on the graph. Now try a β of -0.9. Since nothing can travel faster than the speed of light, an object that begins at the origin is forced to have a worldline between the two lines on the graph. The only object that can have a worldline on either of those lines is light. If we let the woman move in two dimensions her motion would be constrained to move within a cone which is called the lightcone. The cone's boundaries mark the possible worldlines that light can have if it starts at the origin at $t = 0$ m.

2.6   EXPLORING WORLDLINES BY MATCHING

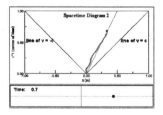

FIGURE 2.8: A spacetime diagram showing the motion of a dragable red ball.

Use the mouse to drag the red ball to change its position. The goal of this Exploration is to match the worldlines shown in the animation (position is given in meters and time is given in the time required for light to travel 1 meter or 3.33×10^{-9} seconds). There is some smoothing of the data sent to the graph. Try to match each worldline, then answer the following questions:

(a) Which of the graphs was the easiest to match and which one was the hardest to match?

(b) Why?

2.7   EXPLORING THE POLE AND BARN PARADOX

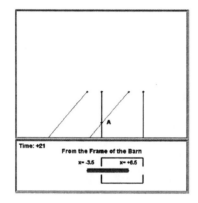

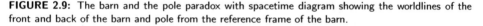

FIGURE 2.9: The barn and the pole paradox with spacetime diagram showing the worldlines of the front and back of the barn and pole from the reference frame of the barn.

A pole vaulter carries her pole towards a barn as shown in the animation (position is given in meters and time is given in the time it takes light to travel one meter or 3.33×10^{-9} seconds). The pole is seen to be 10-m long when it is moving as seen from the frame of the barn. Also shown above the barn is the spacetime diagram depicting the ends of pole and the barn from two frames of reference. When an object is stationary in a reference frame it is red and when it is moving it is green.

Select "View from Barn" and play.

(a) How fast is the pole moving relative to the barn (measured in c)?

(b) What is 1/slope of the red and green worldlines, respectively?

(c) What do events A, B, A′, and B′ refer to? Which if any of these events are simultaneous in this reference frame?

Select "View from Pole" and play.

(d) How fast is the barn moving relative to the pole (measured in c)?

(e) How long is the pole in this reference frame? Why is it this long?

(f) How long is the barn in this reference frame? Why is it this long?

(g) What do events A, B, A′, and B′ refer to? Which if any of these events are simultaneous in this reference frame?

2.8 EXPLORING THE TWIN PARADOX

FIGURE 2.10: The twin paradox with the spacetime diagram showing the worldlines of the traveling and stationary twins.

In this Exploration we will be considering different aspects of the so-called twin paradox. At $t = 0$ years the traveling twin (represented by the green circle) heads out on her journey and then returns at $t = 10$ years (position given in lightyears). In the top panel the spacetime diagram for the stationary frame is shown.

Select "View from Earth-Bound Twin" and play.

(a) How fast is the moving twin traveling relative to the stationary twin (measured in c)?

(b) What is 1/slope of the red and green worldlines, respectively?

Select "Pulses from Earth-Bound Twin" and play. The two twins have agreed to send each other a light pulse once a year on the anniversary of the traveling twin's departure. Also shown is the stationary twin's clock and tick marks on the spacetime diagram depict the sending of the stationary twin's light pulse.

(c) What is the frequency of the stationary twin's light pulse?

(d) How many light pulses reach the moving twin during her outbound trip? During her inbound trip?

(e) At the end of the trip how old is the stationary twin?

One of the most important concepts in special relativity is the idea of the space-time interval. The spacetime interval is $(\Delta s)^2 = c^2(\Delta t)^2 - (\Delta x)^2$ which is the Pythagorean theorem of spacetime.

Select "Pulses from Traveling Twin" and play. Now we have added the traveling twin's clock and numbers on the spacetime diagram to mark the arrival of the traveling twin's light pulse. After you watch this animation, select "Pulses from Traveling Twin: ST" which adds the traveling twin's light pulses to the spacetime diagram.

(f) Calculate the spacetime interval for the stationary twin during the animation.

(g) Calculate the spacetime interval for the traveling twin's outbound trip.

(h) Calculate the spacetime interval for the traveling twin's inbound trip.

(i) Compare the sum of (g) and (h) to (f). Why is there this difference?

PROBLEMS

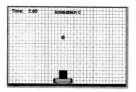

2.1. A ball popper on a cart is shown moving on a track in 4 different animations (position is given in meters and time is given in seconds). In each animation, the ball is ejected straight up by the popper mechanism at $t = 1$ seconds. Consider the reference frames that move along with the carts, in other words, a reference frame in which the velocity of the cart is stationary. Which of these is not an inertial reference frame?

2.2. Three men, A, B, and C, are traveling on a train that travels by you at relativistic speed as shown in the animation (position is given in meters and time is given in seconds). As B passes you, both you and B notice light signals from A and C arriving at the same time. Who emitted the signal first?

2.3. A spaceship flies close to a space beacon at 70% of the speed of light (position is given in meters and time is given in 10^{-5} seconds). The beacon emits light flashes as shown in the animation. What is the time difference between these light flashes as seen by an observer inside the space ship? The time shown in the upper left hand corner is the time as measured in the reference frame of the beacon.

2.4. A pole vaulter carries her pole towards a red barn as shown in the animation (position is given in meters and time is given in seconds). The vaulter carries the pole at a speed such that it is contracted as shown from the reference frame of the barn. What is the length of the pole in its own reference frame?

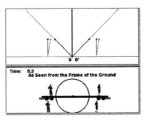

2.5. Shown is an animation of a moving railcar as seen from a stationary observer on the ground. Above this animation is the spacetime diagram for what an observer in the reference frame of the ground would see. Draw the spacetime diagram for what an observer on the railcar would see.

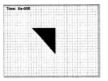

2.6. A triangle is shown traveling to the right as shown in the animation (position is given in meters and time is given in seconds). With what speed must the triangle travel in order to be an isosceles triangle?

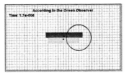

2.7. Two measuring sticks pass close by each other with relative speeds comparable to the speed of light (position is given in meters and time is given in seconds). A light flash occurs when the right ends of the sticks coincide, event 1, and again when the left ends coincide, event 2. Which of the following is true for all relative velocities?

(a) There is a unique speed when the events are simultaneous in both reference frames.

(b) Event 1 always occurs before event 2.

(c) At a given speed, both the red and green observers will always agree on which event occurred first.

(d) None of the above.

C H A P T E R   3

Relativistic Mechanics

3.1 EXPLORING RELATIVISTIC MOMENTUM
3.2 UNDERSTANDING MASS-ENERGY EQUIVALENCE
3.3 UNDERSTANDING THE ENERGY-MOMENTUM EQUATION
3.4 EXPLORING PARTICLE DECAYS
3.5 UNDERSTANDING THE RELATIVISTIC DOPPLER EFFECT
3.6 EXPLORING THE TWIN PARADOX USING THE DOPPLER EFFECT

INTRODUCTION

In introductory physics, we often begin with kinematics and then use those concepts in our development of Newtonian mechanics. We will do something similar here. Now that we understand special relativity and its implications for space and time, we need to examine its implications for mechanics particularly for relativistic momentum and energy. We will also explore measurements of the wavelength and/or frequency of light produced by moving sources (the relativistic Doppler effect).

3.1 EXPLORING RELATIVISTIC MOMENTUM

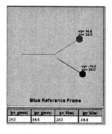

FIGURE 3.1: A relativistic collision between two particles.

Momentum, defined in introductory-level physics as $m\mathbf{v}$, needs a revised definition to account for special relativity. To understand why, consider the animation, which shows you a collision between two objects in a laboratory reference frame and in the reference frame of the blue ball. The switch between reference frames requires a Lorentz transformation from the lab frame to another reference frame (the blue frame). The blue frame is moving at a speed u along the x axis with reference to the lab frame, and the velocity, $\mathbf{v}$, in the lab frame is seen as $\mathbf{v}'$ in the blue frame as follows:

$$v'_x = \frac{(v_x - u)}{(1 - v_x u/c^2)} \ , \quad v'_y = \frac{v_y}{\gamma(1 - v_x u/c^2)} \ , \quad \text{and} \quad v'_z = \frac{v_z}{\gamma(1 - v_x u/c^2)} \ . \quad (3.1)$$

Since the blue frame is moving to the left at a speed of v relative to the lab frame ($u = -v$), the transformation equations give the following velocities for the collision of two balls headed toward each other with speed v in the lab frame:

		Blue Reference Frame	
		Before Collision	After Collision
Green	$v'_{x\ green}$	$2v/(1 + v^2/c^2)$	v
	$v'_{y\ green}$	0	$v(1 - v^2/c^2)^{1/2}$
Blue	$v'_{x\ blue}$	0	v
	$v'_{y\ blue}$	0	$-v(1 - v^2/c^2)^{1/2}$

The animation shows the values for the different incident speeds. If you simply use $\mathbf{p} = m\mathbf{v}$, momentum is not conserved. Momentum conservation is fundamental, so we redefine momentum (and this is the value that is displayed in the table when you check the "show the relativistic momentum" check box):

$$\mathbf{p} = m\mathbf{v}(1 - v^2/c^2)^{1/2} \ , \quad (3.2)$$

where $\mathbf{v}$ is the velocity of the particle as measured in a particular reference frame, not the relative velocity of the reference frame (which is why we do not write v^2/c^2 as β^2; β^2 is instead u^2/c^2). Notice that at low enough speeds, the relativistic momentum and non-relativistic momentum give the same values. It is only for speeds near the speed of light that there is a noticeable difference between the two values.

3.2 UNDERSTANDING MASS-ENERGY EQUIVALENCE

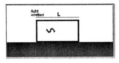

FIGURE 3.2: A massive box in which light has just been emitted from the left wall.

$$E = mc^2 \ . \quad (3.3)$$

This is the one physics equation that almost everyone knows, but most do not understand. The animation shows a *thought experiment* that is generally attributed to Einstein as a way to see the equivalence of mass and energy.[1]

A laser, attached to a box, emits light. The box is on a frictionless surface. The light carries with it energy, and thus, momentum: $p = E/c$. Since the light

---

[1]Based on the development in J. Bernstein, P. M. Fishbane, and S. Gasiorowicz, *Modern Physics*, Prentice Hall (2000), pp. 81–82.

carries a momentum, the box must *recoil*, much like a gun recoils upon firing a bullet. In the animation, therefore, we see the box move. When the light hits the right side of the box, it is absorbed, bringing the box to a stop. What happens to the center of mass? The "center of mass of the box" has clearly moved to the left. Since there are no external forces, the "center of mass of the system" (box plus light) must remain fixed. To keep the center of mass of the system fixed, the light (which travels from left to right) must provide the missing mass. However, light has no mass, just energy. Thus, the energy of the light must be equivalent to the amount of mass needed to keep the center of mass of the system fixed.

Quantitatively, we can compare the distance the box moved with the center of mass of the light motion. Assuming the box is massive enough that its motion is non-relativistic, the amount the box moves is determined from conservation of momentum:

$$p_{\text{box}} = -p_{\text{light}} = -E/c \,, \tag{3.4}$$

which implies $v_{box} = E/Mc$ where M is the mass of the box.

When the light hits the left side of the box, and is absorbed, the box stops. The time that it takes the light to hit is *approximately* L/c (it is actually shorter because the box is moving toward the light, but for our purposes, we will neglect this small effect). The distance the box moves is

$$d_{\text{box}} = v_{\text{box}}t = EL/Mc^2 \,. \tag{3.5}$$

For the center of mass to stay fixed, the light, which traveled a distance L, must have a *mass equivalent* given by the following:

$$Md_{\text{box}} = m_{\text{equiv light}}L \,, \tag{3.6}$$

or

$$MEL/Mc^2 = m_{\text{equiv light}}L \,, \tag{3.7}$$

and hence, $m_{\text{equiv light}} = E/c^2$, the equivalence of mass and energy!

3.3   UNDERSTANDING THE ENERGY-MOMENTUM EQUATION

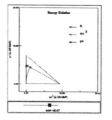

FIGURE 3.3: A graphical depiction of the relationship between total energy, momentum, and mass in special relativity.

The animation shows the relationship between total energy, momentum, and mass (energy is in MeV, mass in MeV/c^2 and momentum in MeV/c). The slider controls the velocity of the object. Set the particle mass and total energy and move the slider. The total energy, E, is given by the equation

$$E^2 = (pc)^2 + (m_0 c^2)^2 = p^2 c^2 + m_0^2 c^4 \ , \tag{3.8}$$

where p is the relativistic momentum, c is the speed of light, and m_0 is the rest mass of the particle (the mass of the particle measured in its reference frame). In the diagram, the hypotenuse represents the maximum energy while the two legs are the mass and maximum momentum. Using this relationship, and the conservation of total energy and momentum, you can determine masses of unknown particles. In this case, the physics of supercollider experiments.

Not surprisingly, as you increase the particle speed, more of the total energy is kinetic energy (the red arrow) since

$$E = m_0 c^2 + K \ . \tag{3.9}$$

Try setting the values so you have a particle with small mass and large maximum energy. What happens? Try the converse. If you have small enough mass and high enough energy (at speeds very close to c), you can use the extreme relativistic approximation of $E \simeq pc$.

Although E and pc vary from reference frame to reference frame (as does the kinetic energy, K), the quantity $E^2 - p^2 c^2$ is relativistically invariant. This is sometimes convenient in problem solving as it can enable you to switch more easily between the laboratory reference frame to a reference frame where a particle is at rest.

3.4   EXPLORING PARTICLE DECAYS

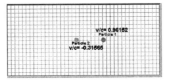

FIGURE 3.4: Two decay products (particles) result from another particle's decay.

The animation shows the decay of one particle into two decay products (momentum is given in units of MeV/c, mass in units of MeV/c^2 and speed is in the form of v/c). If a particle of unknown mass decays into two particles of known mass and momentum, you can discover the mass of that particle by measuring the energies of the end products. The key is to use conservation of relativistic momentum, $\mathbf{p} = m_0 \mathbf{v}(1 - v^2/c^2)^{1/2}$, and energy, $E^2 = p^2 c^2 + m_0^2 c^4$.

Using the animation, choose a mass and an initial speed of a particle that will spontaneously decay into two other particles. Pick the masses of the two decay products that you want to create. If the total mass of the two decay particles is equal to the initial particle mass, check that the momentum is conserved. Compare momentum before and after the decay. When the decay products have different masses, notice that there are two possible scenarios: I) both particles move to the right and II) Particle 2 moves to the left while Particle 1 moves to the right. By reversing the masses of the decay products, you can see both cases (the animation is designed to always show Particle 1 heading off to the right).

Now run the animation in reverse. Notice that you see two particles colliding and creating a third particle that is more massive than the sum of the two initial particles. How is this possible? Where does the additional mass come from?

Note that the animation obeys the laws of relativistic mechanics and thus you cannot create any possible combination of decay products from a given initial particle. For example, the animation will not let you create two decay products of total mass greater than the initial particle's mass. Why not? You might think that if you set the initial particle speed high enough, the additional energy (in the form of kinetic energy) could be converted into rest mass energy. The easiest way to see why this cannot occur is to switch reference frames. Suppose you are in the reference frame of the initial particle. In this reference frame, the initial energy is

$$E_{\text{initial}} = M_{\text{initial}}c^2 \; , \tag{3.10}$$

and it would decay into two particles, each with energy given by the sum of kinetic energy and rest mass energy so

$$E_{\text{decay products}} = K_1 + m_1c^2 + K_2 + m_2c^2 \; . \tag{3.11}$$

For energy to be conserved, the kinetic energy $(K_1 + K_2)$ would need to be negative, but relativistic kinetic energy is given by

$$K = [m_0c^2/\sqrt{1 - v^2/c^2}] - m_0c^2 \tag{3.12}$$

which cannot be negative, any more than non-relativistic kinetic energy can. Since the kinetic energy cannot be negative, it is not possible for the combined masses of the decay products to be greater than the initial particle mass (in all reference frames).

3.5   UNDERSTANDING THE RELATIVISTIC DOPPLER EFFECT

FIGURE 3.5: Three spectral lines from a source as seen when the source is at rest.

In the animation, add a light source (in the visible part of the spectrum). Change the speed of the light source and then choose whether the source is moving toward you or away from you and observe what happens. You can measure the wavelength of the light (given in nm) by clicking in the animation.

Just as you hear different pitches (frequencies) of sound when a car goes past (due to the Doppler effect), light from a moving source is Doppler shifted as well. The Doppler effect for sound waves depends not only on the relative motion between you and the source of sound, but also on the relative motion between the sound source and the air. Initially, scientists thought light had to propagate through some medium which they called the *ether*. This ether provided an absolute reference frame for light. The speed of light would be measured relative to this reference frame, and if you were moving relative to the ether, then you would measure a different speed of light. Special relativity discards this: the speed of light is the same no matter the reference frame. There is no ether and no absolute reference frame. Therefore, the Doppler shift for light is simply due to the relative motion of the two reference frames (the source and receiver) and is given by

$$\frac{1}{\lambda'} = \frac{1}{\lambda} \left[\frac{(1 \mp \beta)}{(1 \pm \beta)} \right]^{1/2} \tag{3.13}$$

for objects moving away from each other (upper signs) and for objects moving towards each other (lower signs). Here $\beta = v/c$, λ is the wavelength as measured in the frame of the moving light source, and λ' is the wavelength as seen in your reference frame.

The Doppler effect provides us with a great deal of information about our universe. For example, light from distant stars is *red-shifted*, that is, the light from the stars is shifted to longer wavelengths (red side of the spectrum). The atoms that make up stars have a unique light spectrum (as we will see later when we explore the nature of atoms) and when we view the spectra from stars, we find the wavelengths to be longer than expected. The wavelength shift means, according to the Doppler effect, the stars are moving away from us. In fact, everywhere we look in the universe, things are moving away from us. This does not mean we are the center of the universe, it simply means that the universe is expanding (think of points on the surface of an expanding balloon: as the balloon expands, all points move away from each other, but no one point is the *center* of the expansion). By measuring the Doppler shift of the wavelength of light we receive from various light sources throughout the universe and knowledge of how far away those sources are, we can calculate how fast the universe is expanding and therefore an approximate age of the universe.

3.6 EXPLORING THE TWIN PARADOX USING THE DOPPLER EFFECT

FIGURE 3.6: A depiction of the twin paradox. Each circle represents a twin and the circles represent the spherical light pulse emitted from the traveling twin.

One twin, Pink, remains on Earth while her twin brother, Green, leaves in a rocket ship to a planet. When he reaches the planet, he turns around at that planet and returns to Earth 20 years later (as measured by Pink, on Earth). Even though Pink is 20 years older, her brother is younger. To measure this, each twin sends out a flash of light on his/her birthday. In his or her reference frame, then, these flashes occur with a frequency of one. What does it look like from the other reference frame? First look at it from Pink's view on Earth. The time clock in the left corner shows the years as measured on Earth (20 years total), while the counter counts the flashes that Pink receives from Green.

(a) What is the frequency of the pulses that she receives from Green on his outbound trip?

(b) What is the frequency of the pulses that she receives from Green on his inbound trip?

Verify that your answers are consistent with the equation for the relativistic Doppler shift, Eq. (3.13), by using the substitution that $1/\lambda = \nu/c$.

Now, from Green's point of view, the situation is a bit different. In order to make this trip, he has to switch reference frames so in the animation, we show the view of the Pink (and Earth) from the two frames: white and black that Green jumps between. In the first reference frame, Pink (and Earth) is moving away from Green, while in the second reference frame, Pink (and Earth) is moving toward him. The pulse counter only counts pulses in the reference frame that Green is currently in.

(c) Verify that the total number of pulses Green receives is equal to Pink's change in age when Green returns to Earth. Notice that Green's clock (in the upper left hand corner) shows him to be younger.

(d) What is the frequency of the pulses that Green receives while in the White reference frame? While in the Black reference frame?

(e) Show that these equations, too, are consistent with the Doppler shift equation, Eq. (3.13).

The Doppler shift only depends on the relative speeds between the two objects. Both Pink and Green measure the same frequency of light from the other person, but Pink and Green receive a different number of signals (Green receives half as many on his trip out as on his return trip) which accounts for the age difference.

PROBLEMS

3.1. The animation shows four particles moving as shown from the yellow particle's reference frame (distances are given in light years, time is given in years and mass is given in MeV/c^2). From yellow's point of view, rank the particles by kinetic energy from smallest to largest (indicate any ties).

3.2. Four graphs are shown representing the relativistic behavior of an electron. The slider controls the velocity of the electron (velocity is given in 10^8 m/s and energy is given in MeV). One graph correctly shows the total energy of an electron; which one is it?

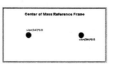

3.3. A Σ^+ (Sigma) particle decays into a neutron and π^+ (pion). What is the mass of the Σ^+? For an initial momentum of 5000 MeV/c of the Σ^+, what is the kinetic energy? The rest mass of neutron is 940 MeV/c^2 and the rest mass of a π^+ is 140 MeV/c^2 (momentum is in units of MeV/c).

3.4. Pick a reference frame to view the creation of an antiproton from a collision of two protons. When the two protons collide, they create three protons and one antiproton (same mass as proton, but opposite sign). The rest mass of a proton (and anti-proton) is 938 MeV/c^2. In the laboratory frame, the kinetic energy of the incoming particle (red) on the target (blue) is called the threshold kinetic energy to produce this reaction.

   (a) In the laboratory frame, what is v/c for the red particle?

   (b) What is v/c for the reaction products (the yellow and green particles)?

   (c) What is the kinetic energy of the red particle (the threshold energy)? The Bevatron accelerator at the Lawrence Berkeley Laboratory was designed to give a proton this threshold kinetic energy to produce this reaction. Emilio Serge and Owen Chamberlain won the 1959 Nobel prize in physics for carrying out this experiment and their discovery of the antiproton as a result.[2]

---

[2]See discussion in K. Krane, *Modern Physics*, John Wiley and Sons (1996), pp. 50-51.

3.5. An unknown particle decays into two photons with wavelengths given in the animation.[3] You can change the incident speed of the unknown particle. You can use the pink protractor to measure angles: click-drag on the gray circles to move and adjust the angle of the protractor. (Angles are given in degrees and wavelengths in nm.) A relationship you may find useful is $hc = 1240$ eV-nm.
  (a) Develop a relationship for the angle the photons leave as a function of $\beta = v/c$. Verify your relationship using the animation.
  (b) Find the mass (in MeV/c^2) of the unknown particle. (What is it?)

3.6. A photon hits an electron at rest as shown in the animation. This generates a positron-electron pair which moves off to the right with the same speed as the electron. A positron has the same rest mass as an electron, but is positively charged. Electron (and positron) rest mass is 0.511 MeV/c^2. What is the momentum of each of the end products? What was the energy of initial photon?

3.7. One twin, Pink, remains on Earth while her twin brother, Green, travels the universe in a fast rocket ship. The animation shows everything from Pink's reference frame (distances are given in light years and time is given in years). Pink sends out light pulses at a given frequency. For each animation,
  (a) What is the frequency of Pink's pulses in Pink's reference frame?
  (b) What is the frequency of the pulses in Green's reference frame?

3.8. The headlights of your physics professor's yellow Lamborghini are sodium vapor lamps. You can measure the wavelength of the light (in nm) by clicking in the animation. When you stand in front of the car "as it sits still," the lights appear to be a bright yellow color.
  (a) With what speed is the car moving if you observe the "spectra from the headlights of the moving car"?
  (b) In which direction (relative to you) is the car moving?

---

[3]Based on Problems 2.37 and 2.39, J. Taylor, C. Zafiratos, and M. Dubson, *Modern Physics*, Prentice Hall (2004).

PART TWO

THE NEED FOR A QUANTUM THEORY

CHAPTER 4

From Blackbody to Bohr

4.1 BLACKBODY RADIATION

4.2 EXPLORING WIEN'S DISPLACEMENT LAW

4.3 BROWNIAN MOTION

4.4 EXPLORING THE *e/m* EXPERIMENT

4.5 EXPLORING THE MILLIKAN OIL DROP EXPERIMENT

4.6 THOMSON MODEL OF THE ATOM

4.7 EXPLORING RUTHERFORD SCATTERING

4.8 THE BOHR ATOM AND ATOMIC SPECTRA

4.9 EXPLORING THE EMISSION SPECTRUM OF ATOMIC HYDROGEN

INTRODUCTION

The late 1800s and early 1900s moved physics from classical physics (described by Newton's laws and Maxwell's equations) to what we now call modern physics, explained by special relativity and quantum theory. The move to a quantum theory was prompted by a series of experiments that could not be adequately explained classically. They include blackbody radiation and a number of other experiments that placed the atom on center stage. This period marks the beginning of our current understanding of the fundamental structure of matter. This chapter ends with the Bohr model of the atom or the so-called *old quantum physics*, which was amazingly successful at explaining a number of atomic properties, but the observations these models could not explain hinted at the need for further explanations (quantum mechanics).

4.1 BLACKBODY RADIATION

One of the first failures of classical theory came about during the analysis of radiation from opaque, or black, objects. Such black bodies radiated and had energy densities, energy per volume per wavelength, $u(\lambda)$, that depended on their temperature. The Rayleigh-Jeans formula for blackbody radiation, derived from the classical equipartition of energy theorem, gives the following functional form for such an energy density:

$$u(\lambda) = 8\pi\lambda^{-4}k_B T \,, \tag{4.1}$$

where $k_B = 1.381 \times 10^{-23}$ J/K is Boltzmann's constant and T is the temperature in Kelvin. Select the "R - J" button on the animation and change the temperature to see how this curve varies. Note the units on the graph (J/m^4 vs. microns) and that the graph's scale changes as you change the temperature. The Rayleigh-Jeans

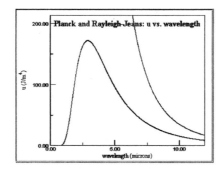

FIGURE 4.1: The Rayleigh-Jeans and Planck blackbody radiation curves.

formula agrees well with the experimental results for very long wavelengths (at low frequencies). As the wavelength of the radiation gets smaller (at high frequencies), the Rayleigh-Jeans formula states that the energy density of the radiation approaches infinity. This does not agree with experiment, however, and the failure of this classical result to agree with experiment is called the *ultraviolet catastrophe.*

Planck solved this problem by treating energy not as continuously variable, but instead, as if it came in discrete units, E_γ, and for light, energy was proportional to frequency

$$E_\gamma = h\nu = hc/\lambda \,, \tag{4.2}$$

where $h = 6.626 \times 10^{-34}$ J·s was a new constant, now called Planck's constant, tuned to fit the blackbody radiation data. When Planck did the $u(\lambda)$ calculation with this assumption, he found:[1]

$$u(\lambda) = \frac{8\pi hc\lambda^{-5}}{e^{hc/\lambda k_B T} - 1} \,, \tag{4.3}$$

which agrees with the experimental data. Select the "Planck" button on the animation and change the temperature to see how this curve varies with temperature. In addition, Wien's displacement law, $\lambda_{\max}T = 2.898 \times 10^{-3}$ m·K, for the wavelength corresponding to the maximum energy density per wavelength, can be verified by looking at the Planck curve.

Because of the agreement between the data and the Planck blackbody radiation law, selecting the "Planck and R - J" button shows just how poorly the Rayleigh-Jeans formula does in replicating the true blackbody radiation curve in the small-wavelength limit.[2]

---

[1]For more details on the derivation of Planck's blackbody radiation law, see P. A. Tipler and R. A. Llewellyn, *Modern Physics* 3ed, W. H. Freeman and Company (1999) or see Section 15.5.

[2]You may also see the Rayleigh-Jeans and Planck formulas in terms of frequency as $u(\nu) = 8\pi\nu^2 c^{-3} k_B T$ and $u(\nu) = 8\pi h\nu^3 c^{-3}/(e^{h\nu/k_B T} - 1)$, respectively. Note that the difference in form is not just due to the substitution $\nu = c/\lambda$. Also note that because of this difference, the graphs of $u(\nu)$ vs. ν will look *flipped around* as compared to $u(\lambda)$ vs. λ.

4.2   EXPLORING WIEN'S DISPLACEMENT LAW

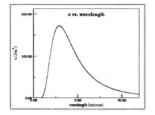

FIGURE 4.2: A blackbody radiation curve, where the peak wavelength corresponds to the maximum energy density per wavelength, $\lambda_{\max}$.

Planck's blackbody radiation law, $u(\lambda) = 8\pi hc\lambda^{-5}/(e^{hc/\lambda k_B T} - 1)$, describes how the energy density per wavelength, $u(\lambda)$, varies with wavelength. You may have noticed that for a given temperature, T, this curve has a maximum energy density per wavelength at a given wavelength. This wavelength, $\lambda_{\max}$, depends on the temperature given by the relationship $\lambda_{\max} T = \text{constant}$. The relationship between $\lambda_{\max}$ and temperature is called Wien's displacement law.

Vary the temperature using the text box and the "set temperature" button to verify Wien's displacement law. What is the constant in Wien's displacement law? Hint: Plot $\lambda_{\max}$ (in meters) versus $1/T$ and determine the slope of this graph.

4.3   BROWNIAN MOTION

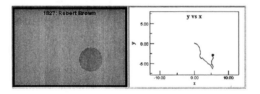

FIGURE 4.3: A large circle representing a grain of pollen moves erratically as seen in Brownian motion. Alongside the animation is a graph of position (x and y) vs. time.

In 1827 botanist Robert Brown noticed that grains of pollen suspended in a liquid moved erratically. This motion is called Brownian motion. Click on the "Brown" animation to see his observations. The graph shows the path of the pollen grain. In 1905 Einstein explained this motion by creating a mental picture or thought (Gedanken) experiment that involved the invisible molecules in the liquid colliding into the larger pollen molecules. The "Einstein" animation shows the *hidden* molecules. Einstein's explanation allowed for a calculation of the number of atoms in a mole (Avogadro's number) and put *atomicity* on solid footing.

In the "Einstein" animation, the smaller (and invisible in optical microscopes) molecules have a kinetic energy that is proportional to their temperature (temperature is a measure of average kinetic energy). These particles collide into the larger particle and cause it to move. Einstein's relationship for the average displacement,

D_{rms}, of a spherical particle of radius r is given by

$$D_{\mathrm{rms}} = \sqrt{k_B T t / \pi \eta r} \,, \qquad (4.4)$$

where k_B is the Boltzmann constant, T is the temperature, t is the time, and η is the viscosity of the fluid in which the particle is suspended. All of the quantities could be measured except k_B, the Boltzmann constant. Determining k_B, however, gave a value for N_A, Avogadro's number since $N_A = R/k_B$ where R is the gas constant (as in the ideal gas law: $pV = nRT = Nk_BT$). Jean-Baptiste Perrin used this method in conjunction with two other methods (one of which was based on the rotation of suspended particles exhibiting Brownian motion, which was also explained by Einstein) to determine Avogadro's number. Perrin won the Nobel Prize in physics in 1926 for this series of measurements that firmly established the idea of atoms as fundamental constituents of nature.

4.4 EXPLORING THE e/m EXPERIMENT

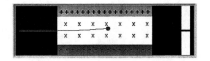

FIGURE 4.4: A simulation of J. J. Thomson's e/m experiment using *crossed* electric and magnetic fields.

The animation illustrates the deflection of an electron in external fields which simulates J. J. Thomson's experiment with cathode rays (position is given in mm, 10^{-3} m). You can change the value of the electric field and play it with the magnetic field, of 2 mT, on or off. After determining that a cathode ray beam was negatively charged, Thomson performed several experiments to further determine the properties of these cathode ray beams, made up of what we now call electrons. By knowing the value of the force on the particles and their initial speed, he reasoned he could determine the mass/charge of the particles.

First, Thomson determined the speed of the rays in the cathode ray tube by passing the beam through crossed electric and magnetic fields. The force on a charged particle with charge q is given by the Lorentz force, $\mathbf{F} = -q(\mathbf{E} + \mathbf{v} \times \mathbf{B})$, where $\mathbf{v}$ is the velocity of the charge, and $\mathbf{E}$ and $\mathbf{B}$ are the electric and magnetic fields, respectively. In this animation, when the magnetic field is off, the particle deflects upward because of the electric field created by the charges shown on the plates ("+" and "-" charges on the blue and green plates). Given a velocity to the right in what direction should the magnetic field be to deflect the particles downward? Once the magnetic field is on, if the electric field is set correctly, the beam does not deflect. Try different values of the electric field until there is no longer a deflection. From this, determine the velocity of the particle. Note that the analysis was all done non-relativistically. Is the speed slow enough to justify this?

Once you have calculated the velocity, turn off the magnetic field and measure the deflection of the beam. Measure the distance the particle deflects in the electric field (the y distance from its entry into the field to its exit). The force is given by $\mathbf{F} = q\mathbf{E}$ so the acceleration in the vertical direction is qE/m and the acceleration in the horizontal direction is zero. Therefore, from basic mechanics,

$$y = 1/2(qE/m)t^2 \quad \text{and} \quad x = vt , \qquad (4.5)$$

and since you cannot measure the time in the animation (just as Thomson could not measure the time the electron was in the field), combine the two equations to solve for the deflection of y for electrons traversing a distance x and determine the value of q/m. There is not any way, in this experiment, to get a value for the mass or the charge independently. Why?

4.5   EXPLORING THE MILLIKAN OIL DROP EXPERIMENT

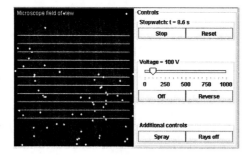

FIGURE 4.5: A simulation of the Millikan oil drop experiment. The little circles represent oil drops with different size and therefore differing amounts of charge. The view is as seen in a microscope.

J. J. Thomson's experiment (Section 4.4) resulted in a value for e/m (the ratio of the charge of the electron to the mass of the electron), but it took the Millikan's oil drop experiment to determine the value of the charge on the electron. This animation is a virtual version of Millikan's experiment.[3] The experiment was based on balancing forces: the downward gravitational force on an oil drop with the upward electric force up on the (ionized) oil drop. Below is a schematic of the apparatus. The sprayer releases ionized oil drops into the apparatus. The oil drops fall and enter a region where they can be seen through the microscope. Turning on the applied voltage provides a force that can, if adjusted correctly, exactly balance the gravitational force on the drop.

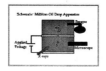

FIGURE 4.6: A schematic of the Millikan oil drop experiment.

---

[3]This Open Source Physics applet was written by Slavo Tuleja.

In the animation, push the "spray" button to spray a group of drops into the virtual apparatus where it is as if you are looking through the microscope (the grid lines are separated by 0.1 mm). Note that each individual drop falls (in the animation, the drops move vertically up because the microscope lens produces an inverted image) at a constant rate (no acceleration). This is because the drops are falling in air where the friction is not negligible and so they reach terminal velocity quickly.

You can "catch" one of the particles by turning on the voltage and adjusting it (push the "On" button and move the slider). The electric field produced (in the real experiment by two plates above and below the view of the microscope) provides a force in the opposite direction to the fall (rise in the animation). Balance a drop somewhere in the screen. Notice that the voltage you use to balance one drop, or several drops, does not balance the rest of the drops. The drops are not all the same size and do not have the same charge (just as in the real experiment).

"Catch" a drop by adjusting the voltage so that it stays still, at least on average. What are the small oscillations due to? (Hint: See Section 4.3.) Balancing the particles, then, you have set the electric force, qE (where q is the charge and E is the magnitude of the electric field) equal to the gravitational force, mg, so that

$$qV/d = mg \,, \tag{4.6}$$

where V is the voltage and d is the distance between the plates applying the voltage. Unfortunately, this still does not provide a value of the charge, q, independent of the mass, m. Record the value of the voltage to suspend this drop. Click the "Rays On" button. The rays you are turning on are X-rays that can change the charge of the drop (the X-rays ionize a drop by providing enough energy for an electron to leave). You will know the drop has changed charge when it starts to move from its *caught location* and head off screen. Now try to adjust the voltage to catch the drop again. Record this new voltage.

In this animation, as in Millikan's experiment, all the drops are not the same size.[4] This means that if you lose the drop you were trying to catch, you have to "spray" out some more drops and catch another one and keep it for a while. Record at least five (5) voltages required to suspend the same drop that you catch. Turn the "Rays On" and "Rays Off" as needed to change the charge of the drop and thus use different voltages to suspend the drop. Looking at the values, can you tell that the charge is quantized? For example, if you were able to record ten different measurements and your data table would look like the following (and it will not for this animation):

Voltage	980	-3030	2950	4060	8020	6970	-2020	5000	1050	6980	-1000

You might think that you could not suspend a drop with a voltage that was not essentially a multiple of 1000 V. If you took many more measurements (as

---

[4]For more details on how Millikan carried out this experiment for measurements on multiple oil drops that were not of uniform size, see J. R. Taylor, C. D. Zafiratos, and Michael A. Dubson, *Modern Physics for Scientists and Engineers*, 2nd ed, Prentice Hall (2004), pp. 108-110 and Problem 3.45, pp. 122-123.

Millikan and his students did), you might argue that the reason that you cannot use a voltage of 1500 V to suspend a charge is that charge comes in discrete units. It is quantized. Look at your measurements for the voltage required to suspend one drop. Do they all (within your errors) have a common factor? What is it? Would you conclude, as Millikan did, that the charge is quantized? Explain.

Once a charge was balanced, Millikan could then find the charge on the electron by turning off the field and determining the terminal velocity of the drop. He did this by timing how long it took the charge to fall a set distance (like across ten grid units in the animation). The terminal speed, v, of a falling drop in a fluid is given by

$$v = 2r^2 \rho g / 9\eta , \tag{4.7}$$

where r is the radius of the drop, ρ is the density of the drop (in Millikan's case it was oil), and η is the viscosity of the fluid (air in this case). By measuring v, he could find the mass of the oil drop since $m = (4/3)\pi r^3 \rho$, and then finally, having m and knowing the electric field, he could calculate the charge.

Catch another drop (or use the same one). Record the voltage necessary to hold it steady. Then, turn the field off, and use the stop watch in the animation to record the time for the particle to fall (to rise as seen in the microscope) across a set number of grid lines. Do this a couple of times with the same drop and determine the terminal velocity of this drop. In this animation, the drops are made out of oil with a density of 875 kg/m^3 and they are falling in air with a viscosity of 7.25×10^{-6} N·s/m^2. The spacing between the plates across which the voltage is applied is 6 mm. Calculate the radius of the drop you caught. From this, determine the charge on the drop. How many excess electrons does it have (or how many electrons have been removed from it)?

4.6   THOMSON MODEL OF THE ATOM

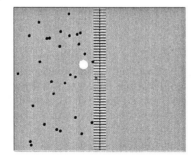

FIGURE 4.7: An alpha particle moving to the right colliding with electrons to simulate alpha particles traveling through a Thomson-like atom.

Once the view of matter as composed of atoms was on sound experimental and theoretical footing, the questioning then focused on the structure of atoms. Atoms were neutral, but contained negatively charged particles, electrons that were too light (5.5×10^{-4} amu) to constitute the mass of the atom (the lightest atom, hydrogen, has a mass of 1 amu). Thomson modeled the atom as made up of a uniformly charged sphere (positive) in which negative electrons were embedded and could move around. The charged sphere had a radius the size of an atom and was massive (in comparison with the electrons). However, when alpha particles (positively charged particles with a mass of 4 amu) hit a thin foil of gold, some of them scattered through very large angles which was not predicted by the Thomson's model.

The animation shows what happens as a massive particle travels through a Thomson-like atom with light electrons. In this model, the particles interact by colliding with each other (hard-sphere collisions) which is not the same as the Coulomb interaction between the electrons and alpha particle. Nevertheless, the overall effect is the same. From conservation of momentum, the more massive alpha particle carries so much more momentum than the light electrons, the electrons barely influence its path. Certainly there is no way to have an alpha particle bounce backwards and in this animation, the yellow particle is only 200 times the mass of the electron instead of the actual 7000 times the mass. The electrons, then, do not noticeably alter the path of the yellow alpha particle.

Maybe, you might argue, instead of electron-alpha particle collisions, the Coulomb interaction between the alpha particle and the uniformly charged sphere could cause scattering. After all, the alpha particle would be repulsed by such a sphere. The animation addresses this possibility.

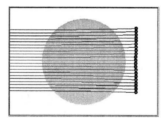

FIGURE 4.8: Twenty alpha particles moving through a positively-charged sphere. The alpha particles do not interact with themselves.

To see the possible deflections, the animation has 20 alpha particles heading toward a positively charged sphere (indicated by the pink region). The alpha particles do not interact with each other, only with the charged sphere. Notice that again, the alpha particles' paths are not noticeably altered.[5]

The Thomson model, therefore, fails to account for an alpha particle hitting an atom and scattering through a large angle. Collisions between the incoming alpha particle and the electrons within the atom, in this model, are like a bowling ball rolling through plastic beads, while the alpha particle traveling through a positively charged sphere is like a bowling ball traveling through water instead

---

[5]See K. Krane, *Modern Physics*, John Wiley, pp. 177-178 for an estimation of the average deflection of an alpha particle by a uniformly positively charged atom. For example, 5 MeV alpha particles scattered from a positively charged sphere with the positive charge of gold ($Z = 79$) and a radius the size of an atom, the average deflection is approximately $0.01°$.

of air. In neither case will the bowling ball recoil. But this is what Rutherford scattering showed: a bowling ball occasionally recoils back towards the original direction of the throw as illustrated in Section 4.7.

4.7   EXPLORING RUTHERFORD SCATTERING

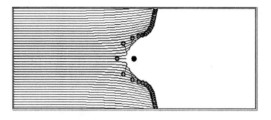

FIGURE 4.9: Twenty alpha particles moving to the right and interacting with a positively-charged nucleus. The alpha particles do not interact with themselves.

The initial animation shows a beam of forty-one alpha particles (test charges) fired at a fixed nucleus. These alpha particles interact via the Coulomb force with the fixed nucleus but not with each other. Because the nucleus is much more massive than any individual alpha particle, modeling this with a fixed nucleus gives a reasonably accurate picture of what happens. Notice that in this case (as opposed to the Thomson model of the atom), the alpha particles can be scattered through large angles if they get close enough to the nucleus and experimentally, alpha particles projected on gold foil do scatter back at large angles.

We can use conservation of momentum and conservation of energy to analyze the scattering. This is an *elastic collision* between a heavy target and a light projectile and the vertical distance the projectile is from the center of the target determines the scattering angle. Because the force is actually the long-range Coulomb force, the *collision* begins well before the particles would touch. Even so, thinking of this like a classical collision works well. Try different initial kinetic energies (and thus different projectile speeds) and target nuclear charge and note how the scattering pattern changes. The animation depicts non-relativistic objects only. For a maximum kinetic energy of 30 MeV, what is the maximum speed of an alpha particle (mass = 3750 MeV/c^2)? Is a non-relativistic approach justified?

The animation can also show one alpha particle impinging on a gold nucleus. You can change the impact parameter (vertical distance between the alpha particle and the nucleus) along with the initial kinetic energy and nuclear charge of the target (the deflection angle is given in degrees). The relationship between the deflection angle, θ, impact parameter, b, and the initial kinetic energy, K, of the incident alpha particle is[6]

$$b = (zZ/2K)(e^2/4\pi\epsilon_0)\cot(\theta/2) , \qquad (4.8)$$

---

[6]See J. R. Taylor, C. D. Zafiratos, M. A. Dubson, *Modern Physics for Scientists and Engineers*, Prentice Hall, 2004, pp. 110-118 or S. Thornton and A. Rex, *Modern Physics for Scientists and Engineers*, 2nd ed, Brooks/Cole, 2002, pp. 123-127 for a full derivation of the Rutherford scattering formula.

where $z = 2$ (charge of alpha particle) and Z is the charge of the nucleus ($Z = 79$ for gold), and $e^2/4\pi\epsilon_0 = 1.44$ eV·nm. Although useful for this animation, what is measured in the lab is the fraction of particles scattered through a given angle as given by the Rutherford scattering formula (derived from the relationship above). The Rutherford scattering formula for the probability per unit area for alpha particles scattering into a ring around the angle θ, called $N(\theta)$ is

$$N(\theta) = (nt/4r^2)(zZ/2K)^2(e^2/4\pi\epsilon_0)^2(1/\sin^4(\theta/2)) , \qquad (4.9)$$

where n is the number of atoms or nuclei per unit volume, t is the thickness of the foil, and r is the distance between the detector and the foil. In the laboratory, the number of particles measured at any given angle followed a $1/\sin^4(\theta/2)$ dependence as predicted by Rutherford's formula. Physicists discarded the Thomson model in favor of the Rutherford model of a positively-charged nucleus that comprised most of the mass of the atom, and is surrounded by orbiting electrons.

4.8   THE BOHR ATOM AND ATOMIC SPECTRA

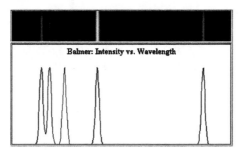

FIGURE 4.10: The visible emission spectra from hydrogen atoms (the Balmer lines).

In 1913 Bohr constructed a model to explain the spectrum of the hydrogen atom. He started with a classical *planetary* model where the electron orbits the nucleus. Elliptical orbits can also be considered, but for simplicity we will use circular orbits.[7] For such a circular orbit, $|F| = ma = mv^2/r = e^2/4\pi\epsilon_0 r^2$, since the electron experiences a Coulomb attraction to the nucleus. Given this result we immediately know that the electrons are accelerating toward the nucleus. Accelerating charged particles radiate classically with the power radiated, $P \propto q^2a^2/c$, where a is the magnitude of the charged particle's acceleration. In fact, they radiate so fast that atoms would exist for only 10^{-10} seconds.

To counter this problem, Bohr postulated that electrons could be found in stable orbits, so-called *stationary states*. These stationary states were characterized by the electron's angular momentum being quantized, $L = nh/2\pi = n\hbar$ ($n =$

---

[7]In fact, elliptical orbits are necessary for the large n limit to agree with the classical prediction for the frequency of the radiation that results from transitions. For large n, transitions require $\Delta n = 2, 3, 4, \ldots$. In the case of circular orbits, the large n limit agrees with the classical prediction for the radiation frequency only for transitions with $\Delta n = 1$.

$1, 2, 3, \ldots$). In addition he postulated that atomic radiation occurred when electrons transitioned from one stationary state to another.

For the hydrogen atom, the total energy is the kinetic plus the potential energy,

$$E = p^2/2m - e^2/4\pi\epsilon_0 r = L^2/2mr^2 - e^2/4\pi\epsilon_0 r = n^2\hbar^2/2mr^2 - e^2/4\pi\epsilon_0 r , \quad (4.10)$$

where we have used Bohr's relationship for L. Re-examining the equation for the force the electron experiences yields a relationship for the radius, $r = e^2/4\pi\epsilon_0 mv^2 = me^2r^2/4\pi\epsilon_0 L^2$ which can be written, using angular momentum quantization, as $r_n = n^2[4\pi\epsilon_0\hbar^2/me^2]$, or n^2 times the Bohr radius (0.53×10^{-10} m). Using r_n in the energy equation yields $E = -me^2/8\pi\epsilon_0 n^2\hbar^2 = -\mathcal{R}/n^2$, where $\mathcal{R} = -13.6$ eV is the Rydberg constant.

Using these results, Bohr could also explain the well-known Balmer lines, the visible emission spectra from the hydrogen atom as shown in the animation. You can click in the animation to measure the wavelength of a given spectral line. Recall that Bohr postulated that atomic radiation occurs when electrons transition from one stationary state to another. Therefore, consider the differences in the energies of two stationary states,

$$E_\gamma = \left(\frac{1}{n^2} - \frac{1}{m^2} \right) \mathcal{R} , \quad (4.11)$$

where n and m are integers with $m > n$. The Balmer lines, as shown in the animation, obey this equation with $n = 2$. For the emission lines, we are seeing light emitted from the atom as it undergoes a transition down to $n = 2$.

Despite this success, there are problems with this model. This simple approach, often called *old quantum theory*, fails to explain the spectra of atoms with more than one electron. It also fails to predict transition rates and spectral line intensities, it gets the electron's angular momentum wrong (for example, in the ground state the electron has zero angular momentum, not $\hbar$), and its use of exact circular orbits with definite angular momentum violates the uncertainly principle (which, to be fair to Bohr, was developed later by Heisenberg).[8]

---

[8]For a nice description of all of the deficiencies of the Bohr model, including the fortuitous two errors he made which canceled, see pages 198-199 of K. Krane, *Modern Physics* 2nd ed. (John Wiley and Sons, New York, 1997).

4.9   EXPLORING THE EMISSION SPECTRUM OF ATOMIC HYDROGEN

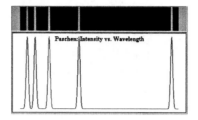

FIGURE 4.11: The infrared emission spectra from hydrogen atoms (the Paschen lines).

The emission spectrum of atomic hydrogen can be considered in several pieces: the Balmer series, which lies mostly in the visible portion of the spectrum; the Paschen series, which lies in the infrared; and the Lyman series in the ultraviolet. The Balmer series can be described by a mathematical formula as described in Section 4.8. Can the Paschen series and the Lyman series be described by a similar mathematical relationship?

(a) Verify that the mathematical formula works for the "Balmer series".

(b) Whereas the leading term in the Balmer formula contains the factor $1/2^2$, the Paschen series and the Lyman series will contain a different factor, but the same form, one over an integer squared. What is the integer for the "Paschen series" and the integer for the "Lyman series"?

(c) Try a different integer in the leading term and calculate the wavelengths of a few lines. Do they match any series?

(d) What combinations of integers produce each series?

(e) Does the same Rydberg constant work for each and every series?

PROBLEMS

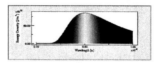

4.1. The graph represents energy density per wavelength emitted from a hot object versus the wavelength of that light. What is the temperature of the object?

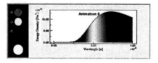

4.2. The graphs represent the energy density per wavelength emitted from various hot objects versus the wavelength of that light. Rank the animations according to temperature (greatest to least).

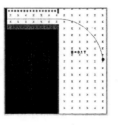

4.3. The animation shows a charged particle traveling first through a region where there are electric and magnetic fields. You can change the value of the electric field (between 0 and 200 kV/m). The magnetic field is 0.1 T in that region as well as the region to the right where there is no electric field (position is given in mm, 10^{-3} m).

(a) What is the velocity of the particle as it passes through the region of crossed electric and magnetic fields?

(b) What is the charge to mass ratio of this particle (it is not an electron)?

4.4. The animation illustrates a charged particle in a Millikan oil drop apparatus (position is given in mm and time in seconds). When the voltage between the plates is off, the drop falls freely with a constant speed because it quickly reaches terminal velocity due to air resistance. When the voltage is on, the droplet is suspended between the plates. Some constants used in this animation that may be useful: density of the oil is 875 kg/m^3, density of air is 1 kg/m^3, and the viscosity of air in this animation is 7×10^{-6} N·s/m^2. The droplet is not shown to scale to make it visible.

(a) If the voltage required to suspend the drop is 280 V, how many excess electrons does the drop have?

(b) If the same drop were to only have one excess electron, what voltage would be required?

4.5. For Rutherford scattering, when an alpha particle ($z = 2$) heads straight for a nucleus (impact parameter, b, equal to zero), the point where the alpha particle gets closest to the nucleus is called the distance of closest approach. You can measure this in the animation above for different values of incident alpha particle kinetic energy and nuclear charge (grid units are in fm, 10^{-15} m).

(a) Develop an expression for that as a function of incident kinetic energy of the alpha particle and charge on the nucleus. Verify your relationship with the animation.

(b) How does this distance compare to the size of the nucleus (around 7 fm for gold)? How do these distances compare to the Bohr radius (typical size of an electron orbital)?

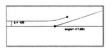

4.6. The animation shows one alpha particle impacting a nucleus. You can change both the incident angle and the impact parameter. What is the charge on the nucleus?

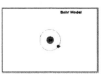

4.7. A Bohr atom with one proton and one electron is shown undergoing a transition as shown in the slow-motion animation (position is given in Bohr radii).
(a) What is the energy of the absorbed light (the green photon)?
(b) What is the wavelength of the absorbed light (the green photon)?

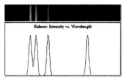

4.8. The spectrum shows part of the Balmer spectrum (light associated with a transition to the $n = 2$ level) for a single electron atom with more than one proton in its nucleus (*i.e.*, it is not hydrogen). What is it (is it singly ionized helium, doubly ionized lithium, triply ionized beryllium)? The units of wavelength on the graphs are nm.

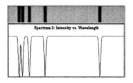

4.9. The animation shows hydrogen absorption spectra. Light shining on an atom is absorbed when the wavelength of light corresponds to an energy difference between two orbitals. Light is absorbed by an electron that uses the energy to make a transition from a lower energy orbital to higher one. The excited states of hydrogen (electrons in $n > 1$) are short-lived states (lifetimes around 10^{-9} s). So hydrogen (at least at reasonable temperatures) is not generally in excited states. Therefore, which of the three spectra above is the absorption spectrum for hydrogen atoms (*i.e.*, which is the absorption spectrum from the $n = 1$ state of the hydrogen atom)? The units of wavelength on the graphs are nm.

C H A P T E R 5

Wave-Particle Duality

5.1 WAVE AND PARTICLE PROPERTIES
5.2 LIGHT AS A PARTICLE: PHOTOELECTRIC AND COMPTON EFFECTS
5.3 EXPLORING THE PROPERTIES OF WAVES
5.4 WAVE DIFFRACTION AND INTERFERENCE
5.5 THE ELECTRON DOUBLE-SLIT EXPERIMENT
5.6 DOUBLE-SLIT EXPERIMENT AND WAVE-PARTICLE DUALITY
5.7 EXPLORING THE DAVISSON-GERMER EXPERIMENT
5.8 DIFFRACTION GRATING AND UNCERTAINTY
5.9 PHASE AND GROUP VELOCITY
5.10 EXPLORING THE UNCERTAINTY PRINCIPLE
5.11 EXPLORING THE DISPERSION OF CLASSICAL WAVES

INTRODUCTION

Waves and particles each have unique properties, often ones that are mutually exclusive. But light, classically considered a wave, sometimes *behaves like* a particle (which we call a photon) and the electron, classically described as a particle, sometimes *behaves like* a wave. This chapter explores the evidence for wave-particle duality of light and electrons (as well as all other *particles*), the implications, and possible ways of understanding this duality. In the process, we also introduce and begin to explore the uncertainty principle: the position and the momentum of a particle cannot both be known precisely at the same time.

5.1 WAVE AND PARTICLE PROPERTIES

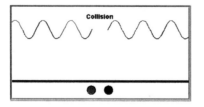

FIGURE 5.1: Two particles and two waves incident on each other just before *collision*.

In the late 1800s, there was a clear division between objects: particles and waves. The resounding success of Maxwell's equations in explaining electricity and magnetism also included a description of light as an electromagnetic wave. In this formulation, light was considered a wave. Consider the following list and identify which of the following describes a wave and which describes a particle:

- Localized.
- More energy, it moves faster.
- One can cancel another one out.
- Collides with one thing at one time.
- Can bend around a corner.
- If two doors are open, part can pass through one and part can pass through the other.

Start the "localization animation." Where exactly is a wave *located*? How can you describe its location at any one point in time? For a traveling sinusoidal wave, it is located everywhere and asking for its location makes no sense. On the other hand, it is easy to identify where the ball is at a given instant of time.

"Increasing the energy" of particles means increasing its speed ($KE = 1/2mv^2$), but for a classical wave of fixed wavelength, an increase in energy (in terms of power delivered) means an increase in amplitude (peak height).

The third and fourth attributes on the list concern what happens when "two objects collide." When waves collide, they add or *superimpose*, while balls bounce off of each other. It is difficult to conceive of two balls colliding into each other and briefly canceling each other out so they disappear altogether. Notice what happens as the waves interact in the animation. There are times when the waves cancel each other out and times when they add to form a bigger wave.

Finally, the last two properties, diffraction and interference, are purely wave phenomena. "Diffraction," is simply the bending of light around a sharp edge: a spreading of the wave front (shown in the ripple tank animation). Interference occurs when waves collide and a good example of when this happens is when a wave passes through a "double-slit (two doors)."

Given these very distinct properties, it was quite surprising when experiments showed that light, although often described as a wave, could not be described solely as a wave. The photoelectric effect and the Compton effect were difficult to explain unless you described light as a particle (Section 5.2). Similarly, although the electron could be described as a point particle, experiments showed that electrons do diffract and interfere (Sections 5.5, 5.6, and 5.7). Thus, there was no clear dividing line between waves and particles. All matter, as it turns out, must be described as both a wave and a particle (complementarity). This property is also called wave-particle duality. How we make a measurement determines whether we will be able to explain the outcome by applying wave or particle properties. If we perform an experiment designed to measure wave-like properties (such as the double-slit experiment), we see wave-like properties (an interference pattern). If we perform an experiment designed to measure particle-like properties (such as the photoelectric effect), we see particle-like properties (light exciting an electron from a metal).

5.2 LIGHT AS A PARTICLE: PHOTOELECTRIC AND COMPTON EFFECTS

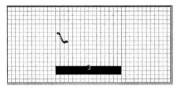

FIGURE 5.2: Light incident on a metal to illustrate the photoelectric effect.

The photoelectric effect, discovered by Hertz in 1887, occurs when light impinges on certain metals and electrons are ejected. These electrons are often called photoelectrons. The kinetic energy of the exiting electrons is measured by applying a stopping potential, V_0, that repels electrons with kinetic energy less than qV_0. In this animation, you may change the incident photon's energy by using the text box and the "enter energy" button. In addition you may change the substance upon which the light impinges by selecting one of the radio buttons.

What was found experimentally was that the kinetic energy of the ejected electrons had a correlation with the frequency (particle behavior) of the light and no correlation with the intensity (wave behavior) of the light. Even with low intensity light, if the frequency is above a certain threshold, ν_0, photoelectrons are emitted from the metal. However, if the frequency is below ν_0, there are no emitted electrons. In 1905, Einstein asserts Planck's hypothesis, $E_\gamma = h\nu$, outside of the realm of the blackbody radiation, postulating that this assumption of Planck was actually a general property of light. Applying it to the photoelectric effect yields

$$E_\gamma = \Phi + eV_0 . \tag{5.1}$$

From Einstein's point of view, quanta of light (photons) were incident on the electrons of a metal and only when the photons had enough energy to overcome the metal's work function, Φ, photoelectrons were produced. Thus, if the light acted like particles of energy $h\nu$, photoelectrons would be produced only if the light had a frequency high enough so that $h\nu$ was bigger than Φ, the work function of the metal.

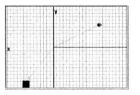

FIGURE 5.3: An electron and a photon after they have interacted via the Compton effect.

Compton used the idea that light behaves like a particle to explain light-electron (photon-electron) scattering. Compton used the relation for the energy,

$E_\gamma = h\nu$, momentum, $p = E/c$ or $p = h/\lambda$, and the relativistic expression for the conservation of energy and momentum, $E_e = \gamma m_e c^2 = (p^2 c^2 + m_e^2 c^4)^{1/2}$, to describe the scattering of light off of an electron.

In this animation, a photon scatters off of an electron, which is initially at rest. You may vary the wavelength of the photon between 0.5 nm and 1.0 nm using the text box and the "fire photon" button. The scattered photon is collected and analyzed at the end of the experiment. Its wavelength and angle from the forward direction are reported. Firing repeatedly allows you to see photons scatter at different angles. How does the wavelength shift depend on the angle of the scattered photon (and the electron)? Observe the change in wavelength of the light. It is given by $\lambda' - \lambda = (h/m_e c)(1 - \cos(\theta))$, where λ' is the scattered photon's wavelength and θ is the photon's scattering angle.[1]

While the work of Einstein and Compton is convincing, treating light as a particle is not the only explanation for the photoelectric and Compton effects. In 1969 Lamb and Scully[2] showed that they could explain the photoelectric effect semi-classically (treating light as a classical electromagnetic wave). Therefore Einstein did not prove the particle-like behavior of light, he only came up with one possible explanation. It was not until 1986 when Grangier, Roger, and Aspect (the so-called Aspect anti-coincidence experiment) experimentally showed that light can indeed behave like a particle. In their experiment, they used a beam splitter and two detectors that were separated by a large distance. For light to display particle-like behavior, the detectors should not detect the same photon. Grangier *et al.* were never able to measure the same photon arriving at the two different detectors. Each photon was measured only at one detector or the other, a perfect anti-coincidence.[3]

5.3   EXPLORING THE PROPERTIES OF WAVES

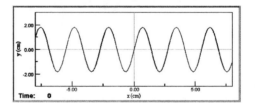

FIGURE 5.4: A traveling wave.

Understanding wave properties includes developing a mathematical relationship for a traveling wave. Shown in black is a traveling wave (position is given in

---

[1]For more details on the derivation of the Compton scattering formula, see Chapter 3 of K. Krane, *Modern Physics*, John Wiley and Sons (1996).

[2]W. E. Lamb, Jr. and M. O. Scully, "The Photoelectric Effect without Photons," in *Polarisation, Matierer et Rayonnement*, Presses University de France (1969).

[3]For more details on this fascinating experiment, see Chapter 2 of G. Greenstein and A. G. Zajonc, *The Quantum Challenge*, Jones and Bartlett (1997); J. J. Thorn, M. S. Neel, V. W. Donato, G. S. Bergreen, R. E. Davies, and M. Beck, "Observing the Quantum Behavior of Light in an Undergraduate Laboratory," *Am. J. Phys.* **72**, 1210-1219 (2004).

centimeters and time is given in seconds). Measure the relevant properties of this wave and determine the wave function of the wave. Once you are finished, check your answer by entering a function, $f(x,t)$, in the text box and looking at the red wave to see if it matches.

With an understanding of the mathematical description of a traveling wave, we can now explore what happens when traveling waves *run into* each other: the superposition of waves. A superposition of two traveling waves is nothing more than the arithmetic sum of the amplitudes of the two underlying waves. We represent the amplitude of a transverse wave by a wave function, $y(x,t)$. Notice that the amplitude, the maximum value of y, is a function of position on the x axis and the time. Given two waves moving in the same medium, we call them $y_1 = f(x,t)$ and $y_2 = g(x,t)$. Their superposition, arithmetic sum, is written as $f(x,t) + g(x,t)$. This may seem like a complicated process, but we simply need to add the waves together at each point on the string (position is given in centimeters and time is given in seconds). Consider the two waves shown above. If these two wave packets are traveling on the same string, draw on a piece of paper the superposition of the two wave packets between $t = 0$ and $t = 20$ s in 2-s intervals for each animation.

When you have completed the exercise, check your answers with the animations provided.

5.4   WAVE DIFFRACTION AND INTERFERENCE

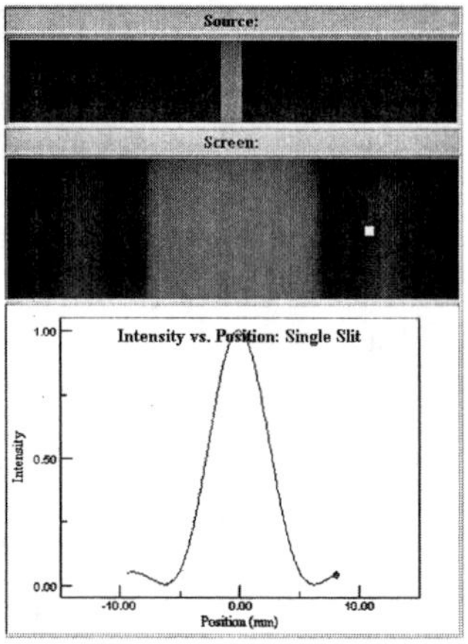

FIGURE 5.5: A single-slit experiment with light. Shown is the source, the resulting projection on a screen, and a graph of intensity vs. position.

When light passes through one slit or two slits, the interference of waves accounts for the observed pattern. First look at light from two distinct sources separated sufficiently far apart to be considered a "double source." What type of pattern do you observe as you drag the detector slowly across the light shown on the screen? (position is given in mm and intensity is in arbitrary units, scaled so that the maximum intensity is 1) How does it change with wavelength (color of light)? How does it change with distance to the screen? Compare this with the "single-slit" and "double-slit" cases. Figure 5.6 shows a schematic of a double slit experiment.

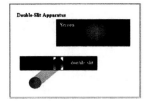

FIGURE 5.6: A schematic of the double-slit experiment.

To qualitatively understand the pattern on the screen in each of the cases, you need to think about the superposition of waves. If two waves *match up* or are *in phase* then the waves constructively interfere and this results in a bright spot on the screen. Light that has traveled the exact same distance from two source points will be in phase and constructively interfere. Constructive interference also occurs when the two light waves travel distances that differ by one, two or other integer multiples of the wavelength. Similarly, if the difference between the two paths is a half-integer multiple of the wavelength of the light, the waves will be out of phase and will destructively interfere (a dark spot). Where are the positions on the screen that correspond to destructive interference?

The quantitative explanation requires further examination of each case. We will begin with the "double-source" case and consider the light on the screen from the first source. It has an electric field given by the real part of $E_1 = E_0 e^{i(kz-\omega t)}$, where k is the wave number $(2p/\lambda)$ and ω the angular frequency. The set up is as shown in Fig 5.7(a). The electric field from the second source takes almost the

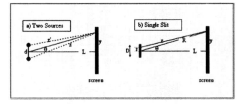

FIGURE 5.7: Geometry for interference from two sources and single slit diffraction.

same form, $E_2 = E_0 e^{i(kz'-\omega t)} = E_0 e^{i(kz-\omega t+\delta)}$, except with a different path length described by the phase shift, $\delta = k(z' - z)$. Thus the total electric field at our point is:

$$E = E_1 + E_2 = E_0 e^{i(kz-\omega t)}(1 + e^{i\delta}) = E_0 e^{i(kz-\omega t)} e^{i\delta}(e^{-i\delta/2} + e^{i\delta/2}), \qquad (5.2)$$

or $E = E_0 e^{i(kz-\omega t)} e^{i\delta}[2\cos(\delta/2)]$.

The intensity of the wave is proportional to $\langle E^2 \rangle$, which is averaged over time. Using the fact that the time average of $\sin^2(\omega t + \phi/2)$ and $\cos^2(\omega t + \phi/2)$

over one period is $1/2$, gives: $I = 2E_0^2 \cos^2(\delta/2)$ or $I_{\text{avg}} = I_{\max} \cos^2(\phi/2)$. When the path length difference is one wavelength, for instance, what is the phase difference between the waves? Since the phase difference, δ, is equal to 2π when the path length difference is one wavelength, λ:

$$\delta = (2\pi/\lambda) * (\text{path length difference}) = (2\pi/\lambda)(d\sin(\theta)) , \qquad (5.3)$$

where θ is the angle the point of interest on the screen is from the center as measured from the source (see diagram). So for small angles, $I = I_{\max} \cos^2(\pi y d/\lambda L)$, where y is the distance along the screen measured from the center and L is the distance between the sources and the screen.

Now, consider a "single slit." We can use the same analysis for one slit, by thinking of the slit as made of multiple sources and the interference between waves from these different tiny sources. In this case, the field due to any small portion of the slit, ds, (see Fig 5.7(b)) is given by $dE = Ae^{i(kz-\omega t)}ds$, where z depends on the value of s. From the diagram (using the law of cosines), $z/R = \sqrt{1 + 2(s/R)\sin(\theta) + (s/R)^2}$, where assuming $R \gg s$, yields $z = R - s\sin(\theta)$. Substituting in this expression for z into dE and integrating from $-D/2$ to $D/2$ where D is the slit width, $E = \int dE = Ae^{i(kR-\omega t)} \int e^{-iks\sin(\theta)} ds$ so the intensity, $I = \langle E^2 \rangle$ is

$$I = I_0 \frac{\sin^2[(kD/2)\sin(\theta)]}{[(kD/2)\sin(\theta)]^2} , \qquad (5.4)$$

and for small angles, $\sin(\theta) \simeq y/L$ and substituting in $k = 2\pi/\lambda$ yields

$$I = I_0 \frac{\sin^2(\pi D y/\lambda L)}{(\pi D y/\lambda L)^2} . \qquad (5.5)$$

Finally, then, the "double slit" combines these two descriptions as the $\cos^2(\alpha)$ interference term (where $\alpha = \pi y d/\lambda L$) falls within an envelope given by the diffraction term of $\sin^2(\beta)/\beta^2$ where $\beta = \pi D y/\lambda L$. Remember that d is the distance between the slits (double-source term) while D is the width of the slit itself (single-slit diffraction term).

5.5   THE ELECTRON DOUBLE-SLIT EXPERIMENT

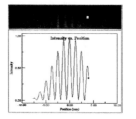

FIGURE 5.8: Sample results from an electron double-slit experiment.

This animation shows double-slit experiments. Light (of a given wavelength) travels through two slits that are close together (close as defined relative to the wavelength of the light). Drag the detector slowly across the light pattern to see the intensity as a function of position (position is given in mm and intensity is in arbitrary units, scaled so that the maximum intensity is 1).

In the second animation, however, electrons are incident on a double slit and pass through to a screen as shown. Each of the five animations gives a snapshot of the data collected at a later time period.

(a) What do you see in the animations?

(b) Do you see an interference pattern evolving?

(c) How does it compare to the interference pattern of light?

(d) Can a single electron end up anywhere on the screen or are there forbidden regions on the screen where an electron cannot land?

Electrons, as it turns out, do create an interference pattern. So even though we think of electrons as particles and when we collide them into other particles they act just as we expect particles to behave, we must, to understand this experiment, think of them as waves that interfere.

5.6   DOUBLE-SLIT EXPERIMENT AND WAVE-PARTICLE DUALITY

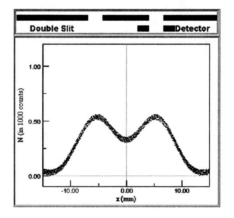

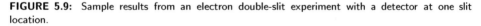

FIGURE 5.9: Sample results from an electron double-slit experiment with a detector at one slit location.

Light acts both like a wave and a particle. Electrons act both like particles and waves. Electrons, in the double-slit experiment, create an interference pattern. This means they are behaving like waves. But are they waves? We know they do not behave like waves when they collide into other objects since, in that case, electrons behave just like particles. Because they act like particles, we tend to think of them as particles (we often have a picture of them as little dots circling the nucleus in planetary-like orbits). But here is an instance of electrons acting like waves.

Let's change the experiment slightly. This time, we will put a detector near one of the two slits to determine which slit the electron goes through. After all, the electron must go through one slit or the other, it cannot go through both, can it?

Look at what happens when you do this experiment. The result is certainly quite different from the original. The interference pattern is gone. It looks as if the signal is clustered directly below each slit. This is exactly what you might expect for a stream of particles to go through one slit or the other. What did we do to get this type of result? We changed the type of measurement that we did. When we try to measure particle-like attributes, that is what we see. When we try to measure wave-like attributes, that is what we see. This is the heart of wave-particle duality.

We have already encountered particle-like properties of light: we've defined momentum (even though light has no rest mass) as $p = h/\lambda$ and energy as $E = pc = hc/\lambda$. What about the wave properties of the electron (or other *particles*)? It turns out that they have a wavelength, called the de Broglie wavelength (Section 5.7) equal to $\lambda = h/p$.

5.7 EXPLORING THE DAVISSON-GERMER EXPERIMENT

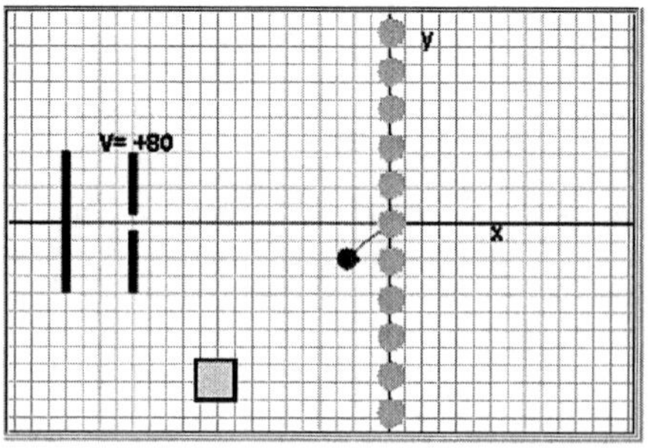

FIGURE 5.10: A simulation of the Davisson-Germer experiment in which electrons are scattered from the atoms in a crystal.

In the Davisson-Germer experiment, electrons with a known initial energy bombarded a surface and the reflection of these electrons from the surface is detected. The surface acts like a diffraction grating with the atoms in the crystal making up the grating.[4]

In the animation, you can change the accelerating voltage, V, or the initial energy of the impacting electrons. The kinetic energy of the electrons is equal to the energy given the electron in the acceleration region. In the animation, the detector moves to the point of maximum signal intensity, the first order diffraction pattern ($n = 1$) (the angle of the detector is given in degrees).

---

[4] As reminder, for light incident on a diffraction grating, the location of a bright spot for a given wavelength, λ, is given by $d\sin(\theta) = n\lambda$, where d is the spacing between the slits in the grating, n is an integer (the order of the maxima), and θ is the angle of the diffracted light from the original path. For commonly-used modern physics experiments that use high energy electrons (keV electrons instead of less than 100-eV electrons) impacting on a material (like graphite), the diffraction pattern is due to interference from electron waves scattered from multiple layers of the material. In that case, the electron beam does not impact the surface at a normal angle and the location of a bright spot is given by the Bragg relationship: $2d\sin(\theta/2) = n\lambda$.

(a) For the range of allowed accelerating voltages $(10 - 100 \text{ V})$, can the electron kinetic energy be considered non-relativistic?

(b) How does the angle of maximum signal depend on the electron energy? Develop a relationship between electron kinetic energy and $\sin(\theta)$.

(c) If the distance between atoms of the material is 0.215 nm (as in nickel), what is the relationship between electron kinetic energy and wavelength?

(d) What then is the relationship of electron momentum to wavelength? You should find that $p = h/\lambda$, which is the de Broglie wavelength.

5.8 DIFFRACTION GRATING AND UNCERTAINTY

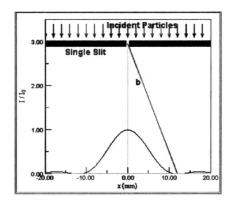

FIGURE 5.11: A single-slit experiment showing the geometry of the problem.

Electrons (or photons) incident on a single slit diffract and form the pattern shown. You can change the slit width (in reality, you cannot control a slit of this width as is necessary for diffracting electrons with such short de Broglie wavelengths in comparison with visible light). Assume the incident electrons are traveling in the y direction as they hit the slit (and have no momentum in the x direction). As they pass through the slit, they are located somewhere within the slit opening of width, D. In order for the beam to spread out, there must now be some momentum in the x direction. What happens to the diffraction pattern as you narrow down the x-position of the particles (by narrowing the slit)? What does this mean about the possible values of the x component of the momentum?[5]

This is a manifestation of the uncertainty principle. The uncertainty in the x direction is about the same size as D (a mathematical analysis shows that this uncertainty is $0.289D$).[6] If we focus only on the central peak in the diffraction pattern, the component of momentum in the x direction must lie between 0 (the maximum intensity) and the amount required to spread to the edge of the central

---

[5] Development follows K. Krane, *Modern Physics*, John Wiley & Sons (1996), p. 118.

[6] The uncertainty in x, Δx, is given by $\Delta x = (\langle x^2 \rangle - \langle x \rangle^2)^{1/2}$ where $\langle x \rangle = \int f(x)\, x\, dx$ evaluated over all space. In this case, $f(x) = 1/D$ for $-D/2 < x < D/2$ and zero for other values of x, thereby ensuring that $\int_{-\infty}^{\infty} f(x)\, dx = 1$.

peak. Specifically, the ratio of the uncertainty in the momentum in the x direction to the momentum in the y direction is given by $\tan(b) = \Delta p_x / p_y$ and, for small angles, $\tan(b) = \sin(b)$. Therefore, $\sin(b) = \Delta p_x / p_y$. However, from our understanding of diffraction from a single slit (Section 5.4), $D \sin(b) = \lambda / 2\pi$, where D is the slit width. Furthermore, the wavelength of the incident electrons is the de Broglie wavelength (see Section 5.7), λ, given by $\lambda = h / p_y$. Combining these equations, we get

$$\Delta p_x = p_y \sin(b) = p_y \lambda / 2\pi D = h / 2\pi D . \qquad (5.6)$$

But, the uncertainty in the x position of these particles is about the slit width ($\Delta x \approx D$) so $\Delta x \Delta p_x \approx h / 2\pi$. This is only an approximate value because the uncertainty in the x value of the momentum is arbitrary (could have been out to the secondary peaks since some small signal appears there as well). However, this is consistent with the uncertainty principle:

$$\Delta x \Delta p_x \approx \hbar . \qquad (5.7)$$

Notice that this is a relationship between the x components. The y component of momentum can be very well known because the particle/wave can exist anywhere in space in the y direction (there is no slit in the y direction).

5.9   PHASE AND GROUP VELOCITY

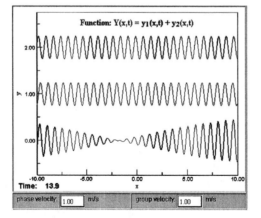

FIGURE 5.12: Two traveling waves and their superposition. The two traveling waves have the same velocity but different wavelengths and therefore frequencies.

A classical wave obeys a classical wave equation. In one dimension the classical wave equation is:

$$\left[\frac{1}{v^2} \frac{d^2}{dt^2} - \frac{d^2}{dx^2} \right] y(x,t) = 0 , \qquad (5.8)$$

where v is the wave velocity and $y(x, t)$ is the wave function, the solution to the wave equation. In general, the solution to this equation is $y(x, t) = f(x \mp vt)$. For harmonic waves, the solutions to the wave equation are

$$y(x, t) = A\cos(kx \mp \omega t) + B\sin(kx \mp \omega t) \,, \tag{5.9}$$

for the right-moving (upper signs) and left-moving (lower signs) solutions.

What do we mean by the velocity of a wave? This may seem like a simple question. When we talk about a wave on a string (or a sound wave) we can talk about the velocity as $v = \lambda f$. We can rewrite this expression in terms of the wave's wave number, k, and angular frequency, ω, given that $\lambda = 2\pi/k$ and that $f = \omega/2\pi$. We therefore find that $v = \omega/k$. We note here that the velocity of the wave is also fundamentally related to the medium in which the wave propagates.

But what happens when we want to add several traveling waves together? In this case we are interested in several waves traveling in the same direction. We can change the wave number and angular frequency for each wave. At first, we will ensure that the wave speeds are identical. In this animation we add the red wave (wave 1) to the green wave (wave 2) to form the resulting blue wave (position is given in meters and time is given in seconds).

Consider what happens when we leave the second wave unchanged, but we change k_1 to 8 rad/m and ω_1 to 8 rad/s for the first wave. Note the interesting pattern that develops in the superposition. Notice that there is an overall wave pattern that modulates a finer-detailed wave pattern. The overall wave pattern is defined by the propagation of a wave envelope with what is called the group velocity. The wave envelope has a wave inside it that has a much shorter wavelength that propagates at what is called the phase velocity. For these values of k and ω the phase and group velocities are the same.

Now consider $k_1 = 8$ rad/m and $\omega_1 = 8.4$ rad/s (now wave one and wave two have different speeds). What happens to the wave envelope now? It does not move! This is reflected in the calculation of the group velocity. The finer-detailed wave has a phase velocity of 1.02 m/s. Now consider $k_1 = 8$ rad/m and $\omega_1 = 8.2$ rad/s. The group velocity is now about half that of the phase velocity (certain water waves have this property). Now consider $k_1 = 8$ rad/m and $\omega_1 = 7.6$ rad/s. The group velocity is now about twice that of the phase velocity (solutions to the free, V = 0, Schrödinger equation have this property).

For a superposition of two waves the group velocity is defined as $v_{\text{group}} = \Delta\omega/\Delta k$ and the phase velocity as $v_{\text{phase}} = \omega_{\text{avg}}/k_{\text{avg}}$. In general, the group velocity is defined as $v_{\text{group}} = \partial\omega/\partial k$ and the phase velocity as $v_{\text{phase}} = \omega/k$.

So what velocity do we want? The physical velocity is that of the wave envelope, the group velocity. For waves on strings we get lucky: the phase and group velocities are the same (these are harmonic waves).

5.10   EXPLORING THE UNCERTAINTY PRINCIPLE

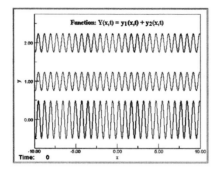

FIGURE 5.13: Two traveling waves and their superposition.

One way to begin to understand wave-particle duality is to think of a particle as a wave packet, constructed from a superposition of waves. Consider the wave given by the equation $y(x, t) = A \cos(kx - \omega t)$ or for one instant in time (picking $t = 0$ for convenience), $y(x) = A \cos(kx)$.

(a) Look at this wave. Where it is *located*?

(b) Keep k_2 the same and choose $k_1 = 8.0$ rad/m. Now, there is a localized packet. What is the uncertainty in x (measure the distance from one zero amplitude to the next)? What is the uncertainty in k ($\Delta k = |k_1 - k_2|$). What is the uncertainty in x and the uncertainty in k for part (a)?

(c) Pick several more values of the wave number, k_1. As the uncertainty in k (Δk) increases, what happens to the uncertainty in x (Δx)?

In general, as the uncertainty in the x position decreases, the uncertainty in the k value increases so that $\Delta x \Delta k \approx 1$. Note that this is just an approximate relationship. In fact, $\Delta x \Delta k \geq 1/2$.

Instead of simply adding two waves together with different k values (and possibly different amplitudes), at a time $t = 0$, we can add a group of waves together. The wave given by the sum (or in the case of a continuous distribution of k values, an integral) of all the superimposed waves is: $y(x) = \int A(k) \cos(kx) \, dk$, where the amplitude of each individual wave added together can depend on the k value. Consider the simplest case where the amplitude is equal to 1 over a range of k values and is zero otherwise.

(d) Integrate the above equation and show that (again this is at $t = 0$): $y(x) = (2/x) \sin(\Delta kx/2) \cos(k_0 x)$ where the amplitude is 1 for $k_0 - \Delta k/2 < k < k_0 + \Delta k/2$.

(e) Use the "Wave Packet: set values" button above to see this wave packet (here, k_0 is the average of k_1 and k_2 and $\Delta k = |k_1 - k_2|$) and measure Δx for different Δk values.

Sections 8.4-8.7 discuss constructing quantum-mechanical wave packets and the time evolution of such wave packets.

5.11   EXPLORING THE DISPERSION OF CLASSICAL WAVES

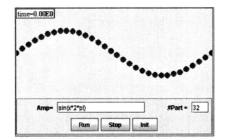

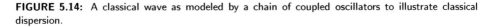

FIGURE 5.14: A classical wave as modeled by a chain of coupled oscillators to illustrate classical dispersion.

The most common example of dispersion is the splitting of white light into a rainbow of color as it passes through a prism. Different wavelengths of light travel though glass at different speeds and so different colors of light are refracted different amounts resulting in a rainbow. This animation allows you to explore dispersion of waves on a string as modeled by a chain of coupled oscillators. You can set the chain in motion by typing in different functions or dragging a red ball before you start the simulation and watch the oscillation of the chain over time (time is given in seconds and position in cm).

(a) Start with the initial equation of sin(2*pi*x) and #Part= 16 and determine the period of the oscillation. What is the speed of the wave? ($v = \omega/k = 2\pi/kT$)

(b) Half the wavelength (Input the equation: sin(4*pi*x) and initialize the animation). What is the period of the oscillation? Speed of the wave?

(c) Continue changing the wavelength and measuring the period and the speed of the wave. How does the wave speed change with the wavelength?

(d) Enter e^(-25*(x-0.5)*(x-0.5)) for the initial condition of a Gaussian wave packet. Describe how its shape changes over time.

Note that this varying wave speed is a variation in the phase velocity of the wave. This is dispersion of a wave: a difference in phase velocity depending on the wave number. This has implications for the time evolution of a wave packet constructed out of a series of these waves. Here the dispersion leads to the change of the wave packet *shape* over time.

PROBLEMS

5.1. In this animation, a slider controls the frequency or wavelength of light incident on a material. A graph shows the kinetic energy of ejected electrons as a function of the slider control (kinetic energy is given in eV).

(a) Animation A. The slider controls the frequency of the incident light (given in 1015 Hz). What is the work function of the material in eV? What is the velocity of the ejected electrons if the incident light has a frequency of $2.5 \times 10^{1}5$ Hz?

(b) Animation B. The slider controls the wavelength of the incident light (given in nanometers, nm). What is the work function of the material in eV?

5.2. The animation shows an apparatus for the photoelectric effect. Light hits the cathode and electrons can be released. You can change the wavelength of light. You can also adjust the voltage that retards the electrons reaching the anode (retarding voltage is given in volts and the photoelectron current is given in nA $= 10^{-9}$ A).

(a) What is the work function of the metal used in the photoelectric effect above?

(b) Using the data for at least three different wavelengths from the animation, determine a value for Planck's constant.

5.3. A 0.14-nm photon incident on a stationary electron scatters as shown in the slow-motion animation. Two depictions of the same event are given.

(a) What is the wavelength of the scattered photon?

(b) What is the magnitude of the velocity of the scattered electron?

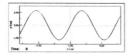

5.4. Shown in black is a traveling wave (position is given in centimeters and time is given in seconds). Measure the relevant properties of this wave and determine the wave function of the wave.

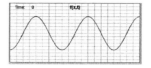

5.5. The animation shows a portion of a standing wave on an idealized taut string (position is given in centimeters and time is given in seconds).
(a) What is the speed of a wave traveling to the right on this string?
(b) Write an equation for the height of the string as a function of both position and time. That is, write a formula for $y(x,t)$.

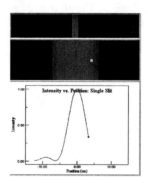

5.6. Drag the detector slowly across the light shown on the screen (in the top panel, slit width and position grid units are in microns (10^{-6} m) while on the graph, position is given in cm and intensity is in arbitrary units, scaled so that the maximum intensity is 1). If the distance to the screen is 20.0 cm, what is the wavelength of the light?

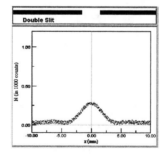

5.7. In this animation, electrons are incident on a double slit and pass through to a screen as shown. Each of the five animations gives a snapshot of the data collected at a later time period. If the slit is 0.08-microns wide and the distance to the screen is 1.2 m, what is the de Broglie wavelength of the electron (signal is in counts of 100 and the position on the screen is given in mm).

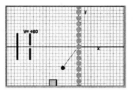

5.8. In the Davisson-Germer experiment, atoms in a crystal lattice form a reflective diffraction grating for the electrons. In the animation, you can change the accelerating voltage, V, or the initial energy of the impacting electrons. The kinetic energy of the electrons is equal to the energy given the electron in the acceleration region. In the animation, the detector moves to the point of maximum signal intensity, the first order diffraction pattern (the angle of the detector is given in degrees). What is the spacing between the atoms?

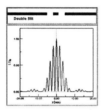

5.9. Photons (or electrons) are incident on a double slit (slit width of 1 micron and separation of 5 microns) and pass through to be detected on a screen as shown in the animation (intensity is given in arbitrary units, scaled so the maximum value is 1 and the position on the screen is given in mm). If wavelength of the incident waves/particles on the double slit is 0.5 microns, how far is the screen from the source? What is the uncertainty of the momentum in the x direction as measured at the screen?

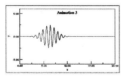

5.10. A 50-gram tennis ball moves to the right as shown in the animation (position given in centimeters and time given in seconds). What is the de Broglie wavelength of the tennis ball?

5.11. Six different classical wave packets are shown in the animations. Which of the wave packets has a phase velocity that is: greater than / less than / equal to the group velocity?

PART THREE

QUANTUM THEORY

CHAPTER 6

Classical and Quantum-mechanical Probability

6.1 PROBABILITY DISTRIBUTIONS AND STATISTICS
6.2 CLASSICAL PROBABILITY DISTRIBUTIONS FOR MOVING PARTICLES
6.3 EXPLORING CLASSICAL PROBABILITY DISTRIBUTIONS
6.4 PROBABILITY AND WAVE FUNCTIONS
6.5 EXPLORING WAVE FUNCTIONS AND PROBABILITY
6.6 WAVE FUNCTIONS AND EXPECTATION VALUES

INTRODUCTION

Much of quantum mechanics involves finding and understanding the solutions to the Schrödinger wave equation and by applying Born's probabilistic interpretation to these solutions.[1] We begin by first reviewing some of the basic properties of classical probability distributions before discussing quantum-mechanical probability and expectation values.[2]

6.1 PROBABILITY DISTRIBUTIONS AND STATISTICS

The animation depicts the result of dropping numerous balls on a peg (or Galton) board. As the individual balls fall, they end up randomly distributed at the bottom of the peg board. The current mean position, position squared, and standard deviation are shown in the table, while a graph of the current mean versus the number of balls dropped is also shown.

In terms of the position of the individual balls that are dropped, x_i, the mean and standard deviation are

$$\langle x \rangle = \sum_{i=1}^{N} \frac{x_i}{N} \quad \text{and} \quad (\Delta x)^2 = \sum_{i=1}^{N} \frac{(x_i - \langle x \rangle)^2}{N}, \tag{6.1}$$

---

[1] While there are at least eight additional formulations, we will primarily focus on Schrödinger's. For all nine, see D. F. Styer, *et al.*, "Nine Formulations of Quantum Mechanics," *Am. J. Phys.* **70**, 288-297 (2002).

[2] This initial focus is suggested in L. Bao and E. Redish, "Understanding Probabilistic Interpretations of Physical Systems: A Prerequisite to Learning Quantum Physics," *Am. J. Phys.* **70**, 210-217 (2002).

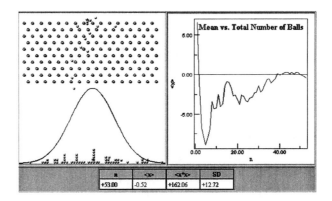

n	⟨x⟩	⟨x²⟩	SD
+53.00	-0.52	+162.06	+12.72

FIGURE 6.1: A Galton or peg board and a graph representing the final position distribution of balls dropped through the peg board.

where N is the total number of individual balls dropped. The standard deviation can also be written in terms of the mean of the position and the mean of the position squared as:

$$(\Delta x)^2 = \langle x^2 \rangle - \langle x \rangle^2 \,, \tag{6.2}$$

since $\sum_{i=1}^{N}(x_i - \langle x \rangle)^2/N$ is the mean of $(x - \langle x \rangle)^2$.

Since there are a finite number of bins (outcomes), as more balls are dropped a pattern emerges from these random outcomes which can be summarized as a discrete probability distribution. On the backdrop of the peg board is a picture of a curve that is a plot of the (continuous) probability distribution function

$$P(x) = \exp\left[-\frac{(x - \langle x \rangle)^2}{2(\Delta x)^2}\right] \,, \tag{6.3}$$

that the discrete distribution will approach for a large number of dropped balls. The functional form of $P(x)$ is called a Gaussian. This is an important functional form for random probability distributions, for classical wave packets, and for quantum-mechanical wave packets.[3]

The Gaussian function is characterized by two parameters: the mean $\langle x \rangle$, which tells us where the peak of the curve falls along the x axis, and the standard deviation Δx, which tells us the width of the curve: the probability distribution drops to $1/e$ of its maximum value at $x = \langle x \rangle \pm \sqrt{2}\Delta x$.

For a continuous distribution, like that of the Gaussian, we no longer sum the individual x_i values. Instead, we integrate the variable of interest over a continuous

---

[3]Note that in the future, when we work with Gaussian wave functions in quantum mechanics, we will be using the normalized form: $\frac{1}{(2\pi(\Delta x)^2)^{1/4}}\exp[-\frac{(x-\langle x \rangle)^2}{(2\Delta x)^2}]$. Since the wave function is the probability amplitude, it is the square of this, $\frac{1}{\sqrt{2\pi(\Delta x)^2}}\exp[-\frac{(x-\langle x \rangle)^2}{2(\Delta x)^2}]$, which is the normalized probability density, $\rho(x)$, in quantum mechanics.

probability distribution that is *normalized*, $P_N(x)$, yielding

$$\langle x \rangle = \int_{-\infty}^{\infty} x\, P_N(x)\, dx \quad \text{and} \quad \langle x^2 \rangle = \int_{-\infty}^{\infty} x^2\, P_N(x)\, dx. \quad (6.4)$$

A continuous probability distribution is normalized when the integral of $P_N(x)\, dx$ over all space is 1.

6.2   CLASSICAL PROBABILITY DISTRIBUTIONS FOR MOVING PARTICLES

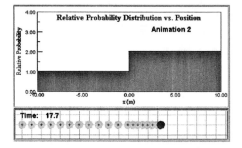

FIGURE 6.2: A classical probability distribution for a particle undergoing a sudden increase in potential energy at $x = 0$.

One of the most important concepts of quantum mechanics is that of a probability distribution in the form of a probability density. Understanding classical probability distributions can help us understand quantum-mechanical probability distributions.

The relative probability distribution, $P_R(x)$, for a classical system can be thought of as the amount of time that a particle spends in a small region of space, $|dx|$, relative to some same-sized region of reference. What is the time spent in a region $|dx|$? Since $|v| = |dx|/dt$, we have that $dt = |dx|/|v|$. The magnitude of the velocity, $|v|$, is related to the particle's total energy, E, via its kinetic energy ($T = mv^2/2 = p^2/2m$) as:

$$|v| = [2(E - V(x))/m]^{1/2}, \quad (6.5)$$

since the kinetic energy of the particle is its total energy minus its potential energy. Therefore the time spent in a region $|dx|$ is

$$dt = \frac{|dx|}{[2(E - V(x))/m]^{1/2}}. \quad (6.6)$$

The relative probability distribution, therefore, is a ratio of the time spent in the region of interest divided by the time spent in the region of reference as long as both regions are the same size, and thus:

$$P_R(x) = \frac{dt}{dt_0} = \frac{|dx|}{[2(E - V(x))/m]^{1/2}} \Big/ \frac{|dx|}{[2(E - V_0)/m]^{1/2}} = \frac{[E - V_0]^{1/2}}{[E - V(x)]^{1/2}}, \quad (6.7)$$

where V_0 is the potential energy in the reference region and $V(x)$ is the potential energy in the region of interest.

Now consider "Animation 1" which depicts a 1-kg particle moving to the right with an initial velocity of 1 m/s. Ghost images are also shown to depict the particle's position at equal time intervals. In the graph above the animation, the relative probability distribution is shaded (position is given in meters and time is given in seconds) so that it is easier to see. What do you notice about it? The relative probability distribution is constant throughout the particle's motion. This occurs because the magnitude of the particle's velocity never changes due to its constant potential energy (recall that $F_x = -dV/dx$, and hence when $dV/dx = 0$, $F_x = 0$ and $a_x = 0$). Therefore, for any same-sized region of space, the time spent is the same as anywhere else.

What about "Animation 2"? In this animation there is a change in the potential energy at $x = 0$ from zero to 3/8 J. Here the particle slows down from a constant 1 m/s for $x < 0$ m to a constant 0.5 m/s for $x > 0$. What happens to $P_R(x)$? There is a distinct change in $P_R(x)$ at $x = 0$ m. If we pick a point with $x < 0$ as our reference point, we note that any point with $x > 0$ has a larger relative probability. Why does this occur? The particle is traveling more slowly at points with $x > 0$ relative to points with $x < 0$, therefore it takes the particle more time to traverse the same distance interval, $|dx|$. Hence the relative probability is greater for $x > 0$ than for $x < 0$.

For bound systems, a particle is confined to move in a finite region of space (between the two classical turning points), and the relative probability distribution can be normalized to yield a normalized probability distribution, $P_N(x)$. This is done by integrating the relative probability distribution over the finite region and then dividing $P_R(x)$ by this result. Because of the normalization, the normalized classical probability distribution now has the unit of one over length. It is $P_N(x) \, dx$ that is interpreted as the probability of finding the particle between x and $x + dx$. Consider "Animation 3" in which a particle is confined to move in a one-dimensional box with infinitely hard walls at $x = -5$ m and $x = 5$ m. The relative probability distribution is uniform from -5 m to 5 m and can be set equal to 1. What is the normalized classical probability distribution? Since $P_R(x)$ is a constant, the integral of $P_R(x)$ from $x_i = -5$ m to $x_f = 5$ m is just 10 m. Therefore $P_N(x) = 0.1$ m^{-1}.

In "Animation 4" the particle is again confined to move in a one-dimensional box with infinitely hard walls at $x = -5$ m and $x = 5$ m. But this time there is a change in the potential energy function at $x = 0$ from $V = 0$ J to $V = 3/8$ J. The relative probability distribution is uniform from -5 m to 0 m and is 1, while from 0 m to 5 m the relative probability distribution is uniform and is 2. What is the classical probability distribution now? We must now be careful when integrating because $P_R(x)$ changes at $x = 0$ m. We get that the integral of $P_R(x)$ from $x_i = -5$ to $x_f = 5$ is 15 m. Therefore for x between -5 m and 0 m, $P_N(x) = 0.066$ m^{-1} and for x between 0 m and 5 m, $P_N(x) = 0.133$ m^{-1}.

6.3 EXPLORING CLASSICAL PROBABILITY DISTRIBUTIONS

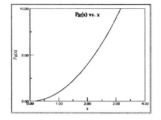

FIGURE 6.3: A possible classical probability distribution.

The relative probability distribution, $P_R(x)$, is shown. For bound systems, where a particle will always be in a certain finite region of space, this distribution can be normalized to yield a normalized probability distribution, $P_N(x)$.

(a) Calculate the relative probability distribution, $P_R(x)$, for a 1-kg particle initially at rest and attached to a spring (a harmonic oscillator) of spring constant $k = 2N/m$ and an initial displacement from equilibrium of 1 m.

(b) Now enter in this function into the text box and set the proper classical limits by clicking the "change limits and evaluate" button. What is the normalized classical probability distribution, $P_N(x)$? You must normalize (a) by doing the integral.

Repeat (a) and (b) for a 0.1-kg particle dropped from rest 1 m above the ground. Assume an elastic collision with the ground at $y = 0$ m.

Once you have completed your calculations, check your results using the animations provided.

6.4 PROBABILITY AND WAVE FUNCTIONS

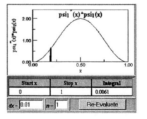

FIGURE 6.4: The probability density of a particle in an infinite square well showing a small slice corresponding to the probability of finding the particle in that small region of space.

The wave function, $\psi(x)$, is a solution to the time-independent Schrödinger equation[4]

$$\left[-\frac{\hbar^2}{2m}\frac{d^2}{dx^2} + V(x) \right] \psi(x) = E\,\psi(x) , \qquad (6.9)$$

in one-dimensional position space (at $t = 0$). The time-independent Schrödinger equation relates the wave function and the energy eigenvalue, E. While we can choose several different variables in which to represent the wave function, like position or momentum, we usually choose position. If we just say, *the wave function*, we almost certainly mean the wave function in position space, $\psi(x)$.

Born's interpretation of the solutions of Eq. (6.9) is that the wave function, $\psi(x)$, represents a probability amplitude at the point x and at a time, t (here $t = 0$). The probability density at the point x is the absolute square of the wave function: $\rho(x) = \psi^*(x)\,\psi(x) = |\psi(x)|^2$. In one-dimension, the probability that a particle between x and $x + dx$ is simply related to the probability density as

$$\rho(x)\,dx = \psi^*(x)\,\psi(x)\,dx = |\psi(x)|^2\,dx . \qquad (6.10)$$

When $\psi(x)$ represents a localized, bound-state solution of the time-independent Schrödinger equation, the integral over the probability density

$$\int_{-\infty}^{\infty} \psi^*(x)\,\psi(x)\,dx = \int_{-\infty}^{\infty} |\psi(x)|^2\,dx = 1 , \qquad (6.11)$$

since the probability of finding a particle somewhere must be one. In order to ensure that the total probability is one, we must often check to make sure that the bound-state wave functions are normalized. Once normalized, wave functions remain normalized for all later times.[5]

The animation shows the probability density for a particle in an infinite square well. The particle is confined between $x = 0$ and $x = 1$. You can change the state, n, and the interval, dx, in which the probability is calculated. Note that there are regions of space in which you would not expect to find the particle. Set dx to 1 and see what happens.

In order to guarantee that the wave function is a solution to the time-independent Schrödinger equation and has a probabilistic interpretation:

- The wave function must tend to a finite value (zero for bound-state solutions and a constant for scattering solutions) as $x \to \pm\infty$. This is a statement of the normalizablility of the wave function. Mathematically, this means that

---

[4]This terminology parallels Styer's [2] terminology that emphasizes

$$\left[-\frac{\hbar^2}{2m}\frac{\partial^2}{\partial x^2} + V(x) \right] \psi(x,t) = i\hbar\,\frac{\partial}{\partial t}\,\psi(x,t) , \qquad (6.8)$$

as *the* one-dimensional Schrödinger equation and Eq. (6.9) as an *energy-eigenvalue* equation, which is a special time-independent case of *the* Schrödinger equation. In general, the Schrödinger equation is three dimensional and has a time dependence. In this chapter we only consider the time-independent case, leaving the time-dependent case to Chapter 7.

[5]This is a property of the Schrödinger equation and the time evolution of states. Quantum-mechanical time development is discussed in Section 7.6.

for bound-state wave functions

$$\int_{-\infty}^{\infty} \psi^*(x)\, \psi(x)\, dx = \int_{-\infty}^{\infty} |\psi(x)|^2 \, dx = 1 \,, \tag{6.12}$$

in order to maintain Born's probabilistic interpretation of ψ.

- The wave function must be continuous and single valued. This means that a valid wave function should not have any *jumps* in it (continuous) and at every point in space have only one value associated with it (single valued). If a wave function was not continuous or not single valued, it would have multiple values for the same position, thereby ruining the probabilistic interpretation of the wave function.

- The wave function must be twice differentiable. In other words, the wave function's first derivative must be continuous, which means that the wave function itself must have no kinks. This is true as long as the potential energy function is itself not severely discontinuous. When the potential energy function has a severe discontinuity, the wave function may have a kink. Examples of such severe discontinuities in the potential energy include the infinite square well (Chapter 10) and the attractive Dirac delta function well (Chapter 11).

6.5 EXPLORING WAVE FUNCTIONS AND PROBABILITY

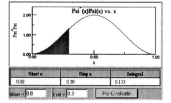

FIGURE 6.5: The probability density of a particle in a one-dimensional infinite square well. The shaded area under the graph represents the probability of finding the particle in that region of space.

The probability density of a particle in a one-dimensional infinite square well is displayed. The probability density is used to determine the probability that the particle's position will lie in a given interval. For the probability density shown in the animation:

(a) Verify that the probability the particle is somewhere, is 1.

(b) Find the probability that the particle is located between $x = 0.30$ and $x = 0.32$.

(c) Find the probability that the particle is located in the left-hand side of the region $(0 \le x \le 0.5)$.

(d) Calculate the expectation value of x—the probability density is given by $2\sin^2(\pi x)$.

(e) Calculate the standard deviation of the distribution.

You may numerically integrate any function with the applet provided.

6.6 WAVE FUNCTIONS AND EXPECTATION VALUES

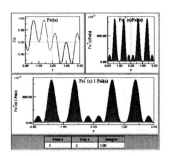

FIGURE 6.6: A wave function, its associated probability density, and the result of the integral of the probability density.

Given a valid wave function, like the one shown in the upper-left-hand corner of the animation, how does the form of the wave function affect the average value or expectation value of variables such as position and momentum? We calculate an expectation value of an observable, $\hat{A}$, as

$$\langle \hat{A} \rangle = \int_{-\infty}^{\infty} \psi^*(x)\, \hat{A}\, \psi(x)\, dx \,, \qquad (6.13)$$

which is the weighted average of $\hat{A}$: the average weighted by the probability density. Explicitly then, if we have a valid wave function, we can calculate the expectation values of

$$\langle \hat{x} \rangle \quad \langle \hat{x}^2 \rangle \quad \langle \hat{p} \rangle \quad \langle \hat{p}^2 \rangle \quad \langle \hat{T} \rangle \quad \langle \hat{V} \rangle \quad \langle \hat{H} \rangle \,, \qquad (6.14)$$

where $\hat{T} = \hat{p}^2/2m$ is the kinetic energy operator, $\hat{V}$ is the potential energy operator, and $\hat{H} = \hat{T} + \hat{V}$ is the Hamiltonian operator.

In general, since the wave function is spread out over space, we expect that there will be some spread (indeterminacy/uncertainty) in our measured values as well. Thus, for position and momentum we can describe this spread using a standard deviation as: $\Delta x = \sqrt{\langle \hat{x}^2 \rangle - \langle \hat{x} \rangle^2}$ and $\Delta p = \sqrt{\langle \hat{p}^2 \rangle - \langle \hat{p} \rangle^2}$ and find

$$\Delta x \Delta p \geq \frac{\hbar}{2} \,, \qquad (6.15)$$

which is the Heisenberg uncertainty principle.[6]

In the animation, a wave function for a particle confined between $x = -2$ and $x = 2$ is shown with its corresponding probability density. In addition you may create the product of a function of x, $f(x)$, and the probability density, and integrate the result. Such a construction will yield $\langle f(x) \rangle$. To make sure that the wave function is normalized, first integrate the probability density by choosing

---

[6] A better description of this concept is actually the phrase 'Heisenberg indeterminacy principle' since an inherent indeterminacy associated with the results of measurements better represents the concept that Heisenberg was trying to describe.

$f(x) = 1$. Now calculate $\langle \hat{x} \rangle$ ($f(x)$ = x) and $\langle \hat{x}^2 \rangle$ ($f(x)$ = x*x). Do your results seem reasonable? The fact that $\langle \hat{x} \rangle = 0$ follows from the form of the probability density (equal probability for $x < 0$ and $x > 0$). However, it may be surprising that $\langle \hat{x}^2 \rangle = 1.2$, given $\langle \hat{x} \rangle = 0$. For $\langle \hat{x} \rangle$, the probability density is weighted by x, which is negative for $x < 0$. For $\langle \hat{x}^2 \rangle$, the probability density is weighted by x^2, which is always positive. We can also calculate the spread in x (standard deviation) and we find that $\Delta x = \sqrt{\langle \hat{x}^2 \rangle} = 1.095$.

PROBLEMS

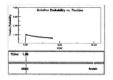

6.1. Five animations depict a ball moving from left to right through the starting and finishing gates. Instead of showing the motion of the ball, a graph is shown of the relative probability (therefore it is not necessarily normalized) of finding the ball near a given position vs. that position. Only when the ball is between $x = 0$ m and $x = 10$ m is this information sent to the graph. View the animations and then:

(a) Rank the animations by the acceleration of the balls. Ties in ( ) please.

(b) Qualitatively describe the potential energy function in each animation.

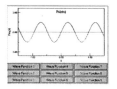

6.2. The animations represent possible bound-state wave functions for well-behaved potential energy functions in one dimension. In the animation, $\hbar = 2m = 1$. Which of the 9 wave functions could be a localized bound-state wave function? Why or why not?

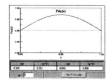

6.3. A particle is in a one-dimensional box of length $L = 1$. The states shown are normalized. The results of the integrals that give $\langle \hat{x} \rangle$, $\langle \hat{x}^2 \rangle$, $\langle \hat{p} \rangle$, and $\langle \hat{p}^2 \rangle$ are also shown. You may vary n from 1 to 10.

(a) For $n = 1$, what are Δx and Δp?

(b) For $n = 10$, what are Δx and Δp?

6.4. A particle is in a one-dimensional harmonic oscillator potential ($\hbar = 2m = \omega = 1$; $k = 2$). The states shown are normalized. Shown are ψ and the results of the integrals that give $\langle \hat{x} \rangle$, $\langle \hat{x}^2 \rangle$, $\langle \hat{p} \rangle$, and $\langle \hat{p}^2 \rangle$. Vary n from 1 to 10.

  (a) When you change n, what do you notice about how $\langle \hat{x} \rangle$, $\langle \hat{x}^2 \rangle$, $\langle \hat{p} \rangle$, and $\langle \hat{p}^2 \rangle$ change?

  (b) Calculate $\Delta x \Delta p$ for $n = 0$. What do you notice considering $\hbar = 1$?

  (c) Calculate $\Delta x \Delta p$ for $n = 10$. How does your result compare to your $n = 0$ result?

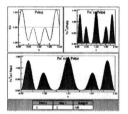

6.5. The wave function, shown in the upper left-hand graph, is for a particle in an infinite square well of length $L = 2$. You may import a function of x and calculate the integral of that function times the probability density. In the animation, $\hbar = 2m = 1$.

  (a) Check that the state is normalized. What $f(x)$ must you use to do this?

  (b) What is the expectation value of x?

  (c) What is the expectation value of x^2?

  (d) What is Δx?

CHAPTER 7

The Schrödinger Equation

7.1 CLASSICAL ENERGY DIAGRAMS
7.2 WAVE FUNCTION SHAPE FOR PIECEWISE-CONSTANT POTENTIALS
7.3 WAVE FUNCTION SHAPE FOR SPATIALLY-VARYING POTENTIALS
7.4 EXPLORING ENERGY EIGENSTATES USING THE SHOOTING METHOD
7.5 EXPLORING ENERGY EIGENSTATES AND POTENTIAL ENERGY
7.6 TIME EVOLUTION
7.7 EXPLORING COMPLEX FUNCTIONS
7.8 EXPLORING EIGENVALUE EQUATIONS

INTRODUCTION

In the previous chapter we began our discussion of quantum mechanics with the wave function and Born's probabilistic interpretation. We now consider the process of how to determine wave functions, the energy of these states, and also how to determine how they evolve in time. To do so, we need look no further than the Schrödinger equation.

7.1 CLASSICAL ENERGY DIAGRAMS

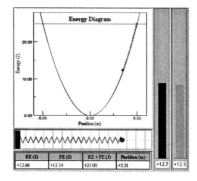

FIGURE 7.1: A classical energy diagram showing total energy and potential energy for a mass on an ideal spring. On the right, bar graphs also represent the potential and kinetic energies.

A large $k = 2$-N/m spring is shown attached to a 1-kg red ball that is initially displaced 5 m (position is given in meters, time is given in seconds, and energy on the bar graph is given in joules). The total energy and the potential energy are shown in the graph. Two bar graphs that depict the kinetic (blue) and potential (green) energies are also shown. Finally, the values of the energy are shown in the table.

The energy diagram is an important diagram for both classical and quantum mechanics because it depicts the potential energy function, often just called the potential. The potential energy function is plotted versus position, and therefore it tells you the potential energy of an object if you know its position. The potential energy function for a mass on a spring is just $V(x) = \frac{1}{2}kx^2$, and therefore in this animation $V(x) = x^2$.

Because of the form of this potential energy function, it is easy to get confused as to what it is actually showing and what it represents. If you have not done so already, run the animation. The red dot on the potential energy curve does NOT represent the actual motion of a particle on a bowl or roller coaster. In other words, it does NOT represent the two-dimensional motion of an object. It represents the one-dimensional motion of an object, here the one-dimensional motion of a mass attached to a spring. The motion of the red mass is limited to the region in between the turning points represented by where the total energy is equal to the potential energy.

Now select the "also show the kinetic energy on the graph" animation. Watch the kinetic and potential energies change as the mass moves and the spring ceases to be stretched and then gets compressed. Notice that sum of the potential energy and the kinetic energy is always equal to the total energy which is a constant. Therefore, if you know the total energy and the potential energy function, you know the kinetic energy of the object at any position in its motion.

7.2   WAVE FUNCTION SHAPE FOR PIECEWISE-CONSTANT POTENTIALS

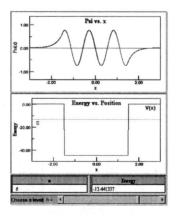

FIGURE 7.2: A position-space wave function for a particle confined to a finite potential energy well.

The animation depicts seven bound states in a *finite square well*.[1] You can use the slider to change the energy level, n, and see the corresponding wave function.

---

[1]This problem is considered in detail in Chapter 11.

We can understand the solution to this problem by analyzing the time-independent Schrödinger equation in one dimension. In regions where the potential energy function does not change too rapidly with position, and can therefore be considered a constant, the time-independent Schrödinger equation is just:

$$\left[-\frac{\hbar^2}{2m}\frac{d^2}{dx^2} + V_0 \right] \psi(x) = E\,\psi(x) \,, \tag{7.1}$$

which we can write as

$$\left[\frac{d^2}{dx^2} - \frac{2mV_0}{\hbar^2} + \frac{2mE}{\hbar^2} \right] \psi(x) = 0 \,. \tag{7.2}$$

In this situation, as in general, there are two cases:[2] $E > V_0$, which is *classically allowed* and $E < V_0$, which is *classically forbidden*. In these two cases the time-independent Schrödinger equation reduces to:

$$\left[\frac{d^2}{dx^2} + k^2 \right] \psi(x) = 0 \quad \rightarrow \quad \psi(x) = A\cos(kx) + B\sin(kx)$$

$$\text{or} \quad \psi(x) = A'e^{ikx} + B'e^{-ikx} \tag{7.3}$$

and

$$\left[\frac{d^2}{dx^2} - \kappa^2 \right] \psi(x) = 0 \quad \rightarrow \quad \psi(x) = Ae^{\kappa x} + Be^{-\kappa x} \,, \tag{7.4}$$

where $k^2 \equiv 2m(E - V_0)/\hbar^2$ and $\kappa^2 \equiv 2m(V_0 - E)/\hbar^2$, so that both k^2 and κ^2 are positive.[3] For an oscillatory solution, the larger the k value, the larger the *curviness* of the wave function at that point[4]

How does this analysis help us understand the wave functions depicted in the animation? In the region where $E > V_0$, the wave function oscillates. In the region that is classically forbidden, $E < V_0$, which corresponds to the far right and far left of the animation, the wave function must be exponentially decaying, $\psi_{\text{left}}(x) \propto e^{\kappa x}$ and $\psi_{\text{right}}(x) \propto e^{-\kappa x}$, in order for the wave function to be normalizable.

---

[2]Usually, the third case $E = V_0$ is not considered in bound-state wave functions except at the classical turning point. There are, however, bound states in which it naturally occurs. For these cases, the time-independent Schrödinger equation becomes: $d^2\psi(x)/dx^2 = 0$, which has a straight-line solution $\psi(x) = Ax + B$. For more examples see Refs. [29, 30, 31].

[3]Even though $E < 0$ and $V_0 < 0$, $E - V_0 > 0$. Thus, $k^2 > 0$.

[4]You may be wondering why we use *curviness* instead of *curvature*. Mathematically, the curvature of a (wave) function is defined by $d^2\psi(x)/dx^2$ which can change *magnitude* and *sign* as a function of position, even when the (wave) function's *curviness* is constant. For example, when $E < V_0$, the curvature of the wave function is such that the wave function curves away from the axis (positive curvature for $\psi(x) > 0$ and negative curvature for $\psi(x) < 0$). For $E > V_0$ the curvature of the wave function is such that the wave function is oscillatory (negative curvature for $\psi(x) > 0$ and positive curvature for $\psi(x) < 0$). Even $\sin(kx)$, which we think of as having a constant *curviness*, has a curvature that depends on position, $-k^2\sin(kx)$. In reality, the only curve that has a constant curvature is a circle.

7.3   WAVE FUNCTION SHAPE FOR SPATIALLY-VARYING POTENTIALS

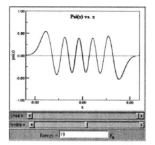

FIGURE 7.3: A position-space wave function for a particle confined to a potential energy well that is proportional to x^2 (a harmonic oscillator).

In one-dimensional position space, the time-independent Schrödinger equation is:

$$\left[-\frac{\hbar^2}{2m}\frac{d^2}{dx^2} + V(x) \right] \psi(x) = E\,\psi(x) \,, \tag{7.5}$$

where $\frac{\hat{p}^2}{2m} = -\frac{\hbar^2}{2m}\frac{d^2}{dx^2}$ since $\hat{p} = -i\hbar\frac{d}{dx}$. To determine the wave function we must solve the time-independent Schrödinger equation for $\psi(x)$ given a $V(x)$.

In the animation a particle is confined to a one-dimensional potential energy well $V(x) \propto x^2$ which describes a *harmonic oscillator potential.*[5] You may change the energy value by varying the sliders (the top slider changes the energy value by $1\,E_0$ while the bottom slider changes the energy value by $0.1\,E_0$) and then examine the solutions to the one-dimensional time-independent Schrödinger equation, Eq. (7.5), for this system. In the animation, $\hbar = 2m = 1$.

The algorithm used to calculate the wave function is called the *shooting method.* The shooting method calculates the wave function by numerically solving the time-independent Schrödinger equation. The solution for the wave function starts with $\psi(x_{\text{left}}) = 0$ to approximate $\psi(x \to \infty) = 0$, and then numerically solves the time-independent Schrödinger equation, calculating the wave function from left to right. Note that all energy values solve the time-independent Schrödinger equation, but only some of these are referred to as *energy eigenvalues* (eigen is German for proper or characteristic) which yield valid (proper) bound-state wave functions. Recall that in order to have a probabilistic interpretation for a bound-state wave function, the wave function must be finite everywhere and must go to zero at $\pm\infty$.

Use both sliders to change the energy to the eigenvalue of $19\,E_0$. As you change the energy, what do you notice about the wave function? The *curviness*[6] changes as the energy changes. As the energy value increases, the *curviness* of the wave function increases. As a consequence, the number of x-axis crossings increases.

---

[5]This problem is discussed in detail in Chapter 12.

[6]We use the term *curviness* much like Robinett [3] uses the term *wiggliness* to qualitatively describe the oscillatory nature of the wave function. Qualitatively, the curviness can be thought of as the number of oscillations per length of the wave function

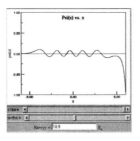

FIGURE 7.4: A possible candidate for the position-space wave function of a harmonic oscillator. This wave function has *too small* an energy to allow the wave function to be finite (in this case zero) at $x = +\infty$.

The ground state always has zero crossings, the first-excited state has one crossing, the second-excited state has two crossings, etc. Now move the lower slider so that the energy value is $18.9\,E_0$. Now increase the energy value to $19.1\,E_0$. What do you notice about what happens to the solution of the time-independent Schrödinger equation? If the energy value is either too small or too large, the wave function either undershoots or overshoots (hence the name *shooting* method) the boundary condition of $\psi(x \to \infty) = 0$ which is approximated by $\psi(x_{\text{right}}) = 0$. When the energy value is too small the *curviness* is too small to allow the wave function to go to zero at $x = x_{\text{right}}$. Likewise, when the energy value is too big, the *curviness* is too large to allow the wave function to go to zero at $x = x_{\text{right}}$.

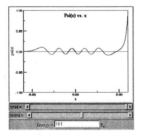

FIGURE 7.5: A possible candidate for the position-space wave function of a harmonic oscillator. This wave function has *too large* an energy to allow the wave function to be finite (in this case zero) at $x = +\infty$.

Also note that for the energy eigenvalue of $19\,E_0$, both the amplitude and *curviness* of the wave function change as a function of x. This is because the potential energy function, $V \propto x^2$, changes with position. As a consequence, $E - V(x)$ changes as well. In classical mechanics, $E - V(x)$ can be easily associated with the kinetic energy, $T(x)$. In quantum mechanics we cannot make this association directly, as it would imply knowing both the position and the momentum at the same time, which violates the uncertainty principle. Instead, we calculate expectation or average values of quantities like the kinetic energy. We may however talk about the *average kinetic energy in a finite interval*, which we refer to qualitatively as curviness.[7] Where $E - V(x)$ is larger, there is a larger curviness in the wave function. Also note that for the energy eigenvalue of $19\,E_0$, you are less likely to find the particle near the origin (since the amplitude of the wave function is related the probability density). For regions where $E - V(x)$ is larger, there is a smaller probability of finding a particle in that region as compared to another region where $E - V(x)$ is smaller. This determination agrees with classical *time spent arguments*[8] for classical probability distributions which was discussed in Section 6.2.

7.4 EXPLORING ENERGY EIGENSTATES USING THE SHOOTING METHOD

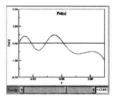

FIGURE 7.6: A possible candidate for a position-space wave function for a particle confined to an unknown potential energy well.

A particle is confined to a box with hard walls at $x = -3$ and $x = 3$ and an unknown potential energy functions within the box. Change the energy slider and examine the solutions to the time-independent Schrödinger equation for this system. In the animation, $\hbar = 2m = 1$.

(a) Determine the energy of the ground state. Start by entering the energy value 4.86 in the slider text box. Use a procedure similar to the following: Enter 4.86. Enter 4.861, 4.862, ... What does the wave function look like when you over/under shoot the energy?

(b) How many energy eigenstates (states that also satisfy the boundary conditions) are between $E = 0$ and $E = 20$?

(c) Determine the energy eigenvalues for the system between $E = 0$ and $E = 20$.

(d) Examine each eigenfunction and sketch a reasonable guess for the potential energy function. Make sure that you also explain your reasoning for the functional form of your sketch.

---

[7]The average kinetic energy can be defined in an interval, ϵ, centered on a point, x_c, as

$$\frac{-\frac{\hbar^2}{2m} \int_{x_c-\epsilon/2}^{x_c+\epsilon/2} \psi^*(x) \frac{d^2\psi(x)}{dx^2}\,dx}{\int_{x_c-\epsilon/2}^{x_c+\epsilon/2} \psi^*(x)\,\psi(x)\,dx}, \tag{7.6}$$

where upon letting $\epsilon \to \infty$ we recover the usual expectation value of the kinetic energy. In quantum mechanics $E - V(x)$, can be negative, and hence the average kinetic energy in a finite interval can be negative. The expectation value of kinetic energy, however, will always be positive.

[8]See for example: R. W. Robinett, *Quantum Mechanics: Classical Results, Modern Systems, and Visualized Examples*, Oxford, New York (1997).

7.5 EXPLORING ENERGY EIGENSTATES AND POTENTIAL ENERGY

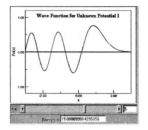

FIGURE 7.7: A possible candidate for a position-space wave function for a particle confined to an unknown potential energy well.

A particle is confined to a box with hard walls at $x = -3$ and $x = 3$ and one of four unknown potential energy functions within the box. Change the energy slider and examine the solutions to the time-independent Schrödinger equation for this system. In the animation, $\hbar = 2m = 1$. Examine each eigenfunction that satisfies the boundary conditions, and sketch a potential energy function that is consistent with your observations.

7.6 TIME EVOLUTION

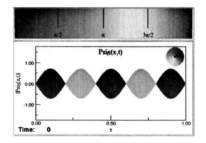

FIGURE 7.8: A time-dependent energy eigenstate of the infinite square well shown in color-as-phase representation.

The Schrödinger equation in one dimension,

$$\left[-\frac{\hbar^2}{2m} \frac{\partial^2}{\partial x^2} + V(x) \right] \psi(x,t) = i\hbar \frac{\partial}{\partial t} \psi(x,t) , \tag{7.7}$$

governs the time evolution of quantum-mechanical systems. We can also write this equation in terms of the Hamiltonian operator, $\hat{\mathcal{H}} = \left[-\frac{\hbar^2}{2m} \frac{\partial^2}{\partial x^2} + V(x) \right]$, as

$$\hat{\mathcal{H}}\psi(x,t) = i\hbar \frac{\partial}{\partial t} \psi(x,t) . \tag{7.8}$$

The left-hand side of the Schrödinger equation involves just the spatial part of $\psi(x,t)$, and for energy eigenstates, we can determine how the spatial part behaves from the time-independent Schrödinger equation. The right-hand side just involves the temporal part of $\psi(x, t)$.

For energy eigenstates, $\hat{\mathcal{H}} \psi(x) = E \psi(x)$, and from the Schrödinger equation the time development of such a state is

$$\psi(x, t) = e^{-\frac{iEt}{\hbar}} \psi(x) , \tag{7.9}$$

while for states in general we have

$$\psi(x, t) = e^{-\frac{i\hat{\mathcal{H}}t}{\hbar}} \psi(x) , \tag{7.10}$$

where $e^{-\frac{i\hat{\mathcal{H}}t}{\hbar}}$ is called the time-evolution operator, $\hat{U}_T(t)$.[9]

Looking at Eq. (7.9) it is clear that time-dependent states will be complex. How do we represent such complex functions? We will demonstrate two possible representations. Consider an arbitrary complex wave function that we choose to write as:

$$\psi(x, t) = \psi_{\mathrm{Re}}(x, t) + i \, \psi_{\mathrm{Im}}(x, t) , \tag{7.11}$$

where we can show the real (blue) and/or imaginary (pink) parts of the wave function. Conversely, when you check the box and "set the state," we can write the wave function as

$$\psi(x, t) = A(x, t) \, e^{i\theta(x, t)} , \tag{7.12}$$

where we have chosen to write all of the complex behavior in the exponential. When we do this, the angle, $\theta(x, t) = \tan^{-1}[\psi_{\mathrm{Im}}(x, t)/\psi_{\mathrm{Re}}(x, t)]$ is called the phase of the wave function. The real part of the wave function in this representation is just the probability amplitude $A(x, t) = \sqrt{\rho(x, t)}$. How do we show the wave function in this amplitude and phase description? We will depict the amplitude of the wave function as the magnitude of the distance from the bottom to the top of the wave function at a given point and time. We represent the phase as the color of the wave function. The color strip above the animation shows the map between phase angle and color. Since quantum-mechanical time evolution involves a minus sign in the exponential, the phase evolves in time *clockwise* in the complex plane.

Explore the time dependence of the energy eigenstate of a particle in an infinite square well by changing state and representation. In the animation the time is given in terms of the time it takes the ground-state wave function to return to its original phase. For energy eigenstates, the amplitude does not change with time, and time evolution just changes the overall phase of the wave function. As a consequence, energy eigenstates are also called *stationary states*.

If there is a time-dependence, we can write the expectation value of an operator as[10]

---

[9]In general, for an operator $\hat{O}$, we know that $\hat{O}^{-1}\hat{O} = \mathcal{I}$. For unitary operators $\hat{O}^\dagger \hat{O} = \mathcal{I}$, and hence, their inverse is their adjoint. The time-evolution operator, $\hat{U}_T$, is a unitary operator because $\exp(i\hat{\mathcal{H}}t/\hbar)^{-1} = \exp(i\hat{\mathcal{H}}t/\hbar)^\dagger$ and therefore $\hat{U}_T^{-1}\hat{U}_T = 1$. This is the case as long as $\hat{\mathcal{H}}$ is a self-adjoint or Hermitian operator, which it must be to yield real energies.

[10]We can think about a situation where the operators have the time-dependence, but the states do not: $\psi_{\mathrm{H}}(x) = e^{i\hat{\mathcal{H}}t/\hbar}\psi_{\mathrm{S}}(x, t)$ and $\hat{A}_{\mathrm{H}}(t) = e^{i\hat{\mathcal{H}}t/\hbar} \hat{A}_{\mathrm{S}} \, e^{-i\hat{\mathcal{H}}t/\hbar}$. This is called the Heisenberg picture. This is equivalent to the situation where the operators have no time dependence but the states do, which is called the Schrödinger picture. The difference between pictures amounts to how you decide to group $\hat{U}_T$ and $\hat{U}_T^\dagger$ in the expectation value: with the states or with the operator.

$$\langle \hat{A} \rangle = \int_{-\infty}^{\infty} \psi^*(x,t)\,\hat{A}\,\psi(x,t)\,dx = \int_{-\infty}^{\infty} \psi^*(x)\,e^{i\hat{\mathcal{H}}t/\hbar}\,\hat{A}\,e^{-i\hat{\mathcal{H}}t/\hbar}\psi(x)\,dx \,, \quad (7.13)$$

and take a time derivative to find:

$$\frac{d\langle \hat{A}\rangle}{dt} = \langle\frac{\partial \hat{A}}{\partial t}\rangle + \frac{i}{\hbar}\langle[\hat{\mathcal{H}},\hat{A}]\rangle\,. \quad (7.14)$$

In general, if a one-dimensional Hamiltonian has a potential $V(x)$ that is just a function of x, we find that:

$$\frac{d\langle \hat{x}\rangle}{dt} = \frac{\langle\hat{p}\rangle}{m}\,, \qquad \frac{d\langle\hat{p}\rangle}{dt} = -\langle\frac{dV(x)}{dx}\rangle\,, \qquad \text{and} \qquad \frac{d\langle\hat{\mathcal{H}}\rangle}{dt} = 0 \quad (7.15)$$

and for the special case of a vanishing potential energy function, we find that:

$$\frac{d\langle\hat{p}\rangle}{dt} = 0\,. \quad (7.16)$$

All of these relationships are described by Ehrenfest's principle, a statement of the quantum-mechanical (on average) and classical correspondence of dynamical variables.

7.7 EXPLORING COMPLEX FUNCTIONS

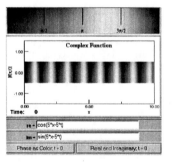

FIGURE 7.9: A quantum-mechanical plane wave shown in phase-as-color representation.

This Exploration allows you to explore complex functions of position and time. You must enter the real and imaginary parts separately. Try the following functions and describe the result in the animation.

(a) re = cos(2*pi*x/10-5*t)   im = sin(2*pi*x/10-5*t)

(b) re = cos(2*pi*x/10)   im = 0

(c) re = 0   im = cos(2*pi*x/10)

(d) re = sin(pi*x/10)*cos(-2*pi*t) im = sin(pi*x/10)*sin(-2*pi*t)

(e) re = cos(3*x-3*t)+cos(4*x-4*t) im = sin(3*x-3*t)+sin(4*x-4*t)

7.8   EXPLORING EIGENVALUE EQUATIONS

FIGURE 7.10: An abstract representation of a state vector (arrow) and the result of an operator acting in the original state vector to produce another state vector (arrow).

The time-independent Schrödinger equation is an energy-eigenvalue equation. What does this mean? Symbolically (this symbolic notation is called Dirac notation) the act of an operator acting on a state can be expressed by

$$\hat{\mathcal{A}}|\alpha\rangle = \beta|\beta\rangle \; , \tag{7.17}$$

where $\hat{\mathcal{A}}$ is an operator, β is typically a number, $|\alpha\rangle$ and $|\beta\rangle$ are two different state vectors. In general, therefore, the action of an operator on a state produces a different state. For special states, however, the final state is the original state and we have

$$\hat{\mathcal{A}}|a\rangle = a|a\rangle \; , \tag{7.18}$$

where $|a\rangle$ is an *eigenvector* or *eigenstate* of the operator $\hat{\mathcal{A}}$ with *eigenvalue* a. This equation is called an eigenvalue equation. To find energy eigenstates of the time-independent Schrödinger equation, we solve $H\psi = E\psi$ *and* solve the boundary conditions.

We can demonstrate the idea of operators, eigenvalues, and eigenvectors by using a 2×2 matrix to represent the operator,

$$\hat{\mathcal{A}} = \left(\begin{array}{cc} \mathcal{A}_{11} & \mathcal{A}_{12} \\ \mathcal{A}_{21} & \mathcal{A}_{22} \end{array} \right) \; , \tag{7.19}$$

and a two-element column vector,

$$|a\rangle = \left(\begin{array}{c} a_1 \\ a_2 \end{array} \right) \; , \tag{7.20}$$

to represent the state vector. An eigenstate of the operator $\hat{\mathcal{A}}$ has the property that the application of $\hat{\mathcal{A}}$ to the eigenvector results in a new vector has the *same direction* as the original eigenvector and a magnitude that is a constant times the eigenvector's original magnitude. In other words: eigenvectors of $\hat{\mathcal{A}}$ are stretched, not rotated, when the operator $\hat{\mathcal{A}}$ is applied to them.

Select a matrix to begin. Drag the red vector around in the animation. The result of the matrix (operator) acting on the column vector (the state vector) is shown as the green vector. To make things easier, you may set the red vector's length equal to 1.

What are the two eigenvectors and eigenvalues of each operator?

PROBLEMS

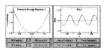

7.1. (a) Which wave function could be an energy eigenstate of which potential energy function? Why?
(b) Which energy level does each wave function correspond to?
Note: The qualitative wave functions could belong to more than one or no potential energy well.

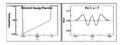

7.2. Shown are a ramped potential energy well and six trial wave functions. You may choose a trial wave function by clicking any "Trial Wave Function" link. Then you may choose a level n from 1 to 7 by dragging the slider to the right to increase n. In the animation, $\hbar = 2m = 1$.
(a) Which Trial Wave Function(s) could represent the energy eigenstates of the green potential energy well?
(b) How do you know? Be as explicit and as complete as possible in your explanation.

7.3. A wave function for a plane wave is shown in two different representations. Shown in the top graph is the amplitude and phase (as color) representation while in the bottom graph the real (blue) and imaginary (pink) components of the wave function are shown separately. In the animation, $\hbar = 2m = 1$.
(a) In the top graph, what color corresponds to a positive and purely real value of the wave function?
(b) In the top graph, what color corresponds to a negative and purely real value of the wave function?
(c) In the top graph, what color corresponds to a positive and purely imaginary value of the wave function?
(d) In the top graph, what color corresponds to a negative and purely imaginary value of the wave function?

7.4. Determine the complex wave function given in the top graph. You may enter the real and imaginary parts of the wave function in the text boxes to check your result.

C H A P T E R 8

The Free Particle

8.1 CLASSICAL FREE PARTICLES AND WAVE PACKETS
8.2 THE QUANTUM-MECHANICAL FREE-PARTICLE SOLUTION
8.3 EXPLORING THE ADDITION OF COMPLEX WAVES
8.4 EXPLORING THE CONSTRUCTION OF A PACKET
8.5 TOWARDS A WAVE PACKET SOLUTION
8.6 THE QUANTUM-MECHANICAL WAVE PACKET SOLUTION
8.7 EXPLORING FOURIER TRANSFORMS BY MATCHING
8.8 EXPLORING WAVE PACKETS WITH CLASSICAL ANALOGIES

INTRODUCTION

Our study of one-dimensional quantum mechanics begins with what you may think is a simple problem: that of a free particle. The Schrödinger equation for this system is as simple as it gets, after all, $V(x) = 0$. However, it is not so simple to construct a *particle-like* solution should we wish to compare quantum mechanics to classical mechanics. In order to do this we must add together an infinite number of individual solutions to the free Schrödinger equation to get a (Gaussian) wave packet that in many ways behaves like a classical free particle.

8.1 CLASSICAL FREE PARTICLES AND WAVE PACKETS

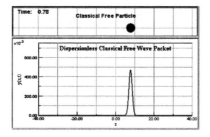

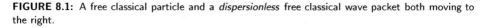

FIGURE 8.1: A free classical particle and a *dispersionless* free classical wave packet both moving to the right.

A classical free particle, like the ball shown in the animation (position given in meters and time given in seconds), obeys the kinematic equation: $x = x_0 + v_0 t$. Classical particles obey Newton's second law, $\sum \mathbf{F} = m\mathbf{a}$, and when there is no net force, there is no acceleration.

A classical wave, such as that of an idealized plucked string,[1] obeys a classical wave equation, $\left[\frac{1}{v^2} \frac{\partial^2}{\partial t^2} - \frac{\partial^2}{\partial x^2} \right] y(x, t) = 0$, where v is the wave velocity and $y(x, t)$ is the wave function. In general, the solution to this equation is in the form $y(x, t) = f(x \mp vt) = g(kx \mp \omega t)$. For harmonic waves, the solutions to the wave equation are

$$y(x, t) = A \cos(kx \mp \omega t) + B \sin(kx \mp \omega t) , \qquad (8.1)$$

where the upper/lower sign describes left-moving/right-moving solutions, respectively. In order to get a localized wave packet, we must add together many such traveling wave solutions. When we do so, we get a wave packet. One such wave packet is a Gaussian wave packet which is shown in the bottom panel of the animation and is described by the wave function

$$y(x, t) = A e^{-(x - x_0 - v_0 t)^2 / \alpha^2} . \qquad (8.2)$$

Here, A is the amplitude, $x_0 + v_0 t$ is the position of the peak of the packet at time t, and v_0 is the packet's velocity, the group velocity. In addition, α describes the width of the packet such that when $x - x_0 - vt = \pm \alpha$, the wave packet drops to $1/e$ of its maximum value. Note that this wave packet maintains its shape throughout its motion. This is because all of the underlying waves that make up this packet have exactly the same velocity since we have assumed harmonic waves with no dispersion.

8.2 THE QUANTUM-MECHANICAL FREE-PARTICLE SOLUTION

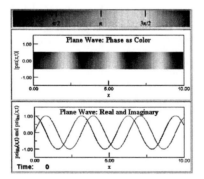

FIGURE 8.2: A quantum-mechanical plane wave in position space. The plane wave is shown in color-as-phase representation and also in the real-and-imaginary components representation. Above the animations is an image that translates color into phase angle.

In order to tackle the free-particle problem, we begin with the Schrödinger equation in one dimension with $V(x) = 0$,

$$-\frac{\hbar^2}{2m} \frac{\partial^2}{\partial x^2} \psi(x, t) = i\hbar \frac{\partial}{\partial t} \psi(x, t) . \qquad (8.3)$$

---

[1]This wave is dispersionless. Waves with dispersion are covered in Section 5.11.

We can simplify the analysis somewhat by performing a separation of variables and therefore considering the time-independent Schrödinger equation:

$$-\frac{\hbar^2}{2m}\frac{d^2}{dx^2}\,\psi(x) = E\,\psi(x)\ . \tag{8.4}$$

which we can rewrite as:

$$\left[\frac{d^2}{dx^2} + k^2\right]\psi(x) = 0\ , \tag{8.5}$$

where $k^2 \equiv 2mE/\hbar^2$. We find the solutions to Eq. (8.5) are of the form $\psi(x) = Ae^{ikx}$ where we allow k to take both *positive* and *negative* values.[2] Unlike a bound-state problem, such as the infinite square well, there are no boundary conditions to restrict the k and therefore E values. In fact, each plane wave has a definite k value and is therefore a definite momentum since $\hat{p}\psi(x) = -i\hbar\frac{d}{dx}\psi(x) = \hbar k\psi(x)$, again with k taking on both positive and negative values (so that $\hat{p}\psi(x) = \pm\hbar|k|\psi(x)$). The time dependence is now straightforward from the Schrödinger equation:

$$i\hbar\frac{\partial}{\partial t}\,\psi(x,t) = E\,\psi(x,t)\ , \tag{8.6}$$

or by acting the time-evolution operator, $\hat{\mathcal{U}}_T(t) = e^{-i\hat{H}t/\hbar}$, on $\psi(x)$. Both procedures yield $\psi(x,t) = A\,e^{ikx-iEt/\hbar}$ (again k can take positive and negative values) and since $E = p^2/2m = \hbar^2k^2/2m$, we also have that

$$\psi_{(k>0)}(x,t) = A\,e^{i(kx-\hbar k^2 t/2m)} \quad \text{or} \quad \psi_{(k<0)}(x,t) = A\,e^{i(-|k|x-\hbar k^2 t/2m)}\ , \tag{8.7}$$

where $\hbar k^2/2m \equiv \omega$. These solutions describe both right-moving ($k > 0$) and left-moving ($k < 0$) plane waves. Recall that solutions to the classical wave equation are in the form of $f(kx \mp \omega t)$ for a wave moving to the right ($-$) or left ($+$). These quantum-mechanical plane waves, however, are complex functions and can be written in the form $f(\pm|k|x - \omega t)$.

In the animation, $\hbar = 2m = 1$. A right-moving plane wave is represented in terms of its amplitude and phase (as color) and also its real, $\cos(kx - \hbar k^2 t/2m)$, and imaginary, $\sin(kx - \hbar k^2 t/2m)$, parts.

What is the velocity of this wave? If this were a classical free particle with non-relativistic velocity, $E = mv^2/2 = p^2/2m$ and $v_{\text{classical}} = p/m$ as expected. But what about our solution? The *velocity* of our wave is ω/k which gives: $\hbar k/2m = p/2m$, half of the expected (classical) velocity! This velocity is the *phase velocity*. If instead we consider the *group velocity*, $v_g = \partial\omega/\partial k$, we find that $v_g = \partial(\hbar k^2/2m)/\partial k = \hbar k/m$, the expected (classical) velocity.

Consider the right-moving wave, $\psi(x,t) = A\,e^{ikx-i\hbar k^2 t/2m}$, which has a definite momentum, $p = \hbar k$. We notice that the amplitude of the wave function, A, is a finite constant over all space. However, we also find that $\int_{-\infty}^{\infty}\psi^*\psi\,dx = \infty$ even though $\psi^*\psi = |A|^2$ is *finite*. While the plane wave is a definite-momentum

---

[2]The most general solution to the differential equation is $\psi(x) = Ae^{ikx} + Be^{-ikx}$ with k values positive.

solution to the Schrödinger equation, it is not a *localized* solution. In this case then, we must discard, or somehow modify, these solutions if we wish a localized *and* normalized free-particle description.[3]

8.3   EXPLORING THE ADDITION OF COMPLEX WAVES

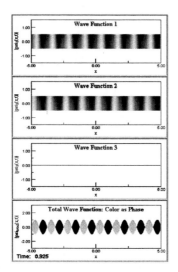

FIGURE 8.3: The addition of two quantum-mechanical plane waves to create a complex standing wave.

In this Exploration we investigate how two and three time-dependent plane waves can be added together to begin to resemble a localized wave packet (in Section 8.4 you can add up to 40 plane waves). In the animation, $\hbar = 2m = 1$.

(a) With the default settings, explain why the arguments of the cosines and sines are of the form (5*x-25*t) and (-5*x-25*t). In other words, what does the ± 5 signify and what does the 25 signify? Remember that $\hbar = 2m = 1$ in this animation.

(b) With the default settings, describe the sum of the two plane waves. Look at the real and imaginary parts of the wave functions to verify your conjecture.

(c) Now change the complex wave functions 2 and 3 to re2 = 1*cos(4*x-16*t), im2 = 1*sin(4*x-16*t), re3 = 1*cos(6*x-36*t), im3 = 1*sin(6*x-36*t). What results? Now change the number multiplying plane wave 2 and plane wave 3 to 0.5. What wave results now? How does this superposition accomplish this result?

---

[3]We can however, *box normalize* the wave function. In box normalization, we normalize the wave function such that over a finite region of space the wave function is normalized.

8.4   EXPLORING THE CONSTRUCTION OF A PACKET

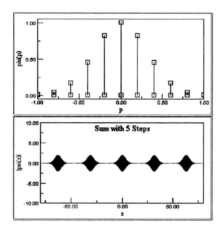

FIGURE 8.4: Individual plane waves added together to form a localized wave packet. Here 11 plane waves are chosen to create the packet and its *copies*. The packet's momentum distribution is also shown.

A localized wave packet can be constructed out of an infinite number of plane wave solutions of the free particle Schrödinger equation. In this Exploration we investigate the process of adding many time-independent plane waves together to resemble a wave packet. We will add these plane waves based on their momentum. We use a Gaussian-shaped weighting function for each plane wave based on its momentum, $\phi(p) = \sqrt{\frac{\alpha}{\sqrt{\pi}}}\, e^{-\alpha^2(p-p_0)^2/2}$. You may choose how many steps, N, are taken in forming the final wave. The larger the N, the larger the number of steps and this amounts to smaller step widths. Pressing the "Add up the Gaussian" button will automatically do this for you. Checking the box below the applet will change the momentum distribution used to construct the packet. In the animation, $\hbar = 2m = 1$.

With the box unchecked and then checked, answer the following questions:

(a) If there are N steps, how many components are there in the sum?

(b) As N increases, what happens to the $\phi(p)$ function?

(c) As N increases, what happens to the position-space wave function? Hint: be systematic. Try $N = 2, 4, 6, 8$, etc. As you do so also use the "set min/max" button to widen your field of view.

(d) If $N \to \infty$, what would the $\phi(p)$ function and position-space wave functions look like?

(e) What are α and p_0 for the two $\phi(p)$ functions?

8.5   TOWARDS A WAVE PACKET SOLUTION

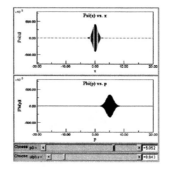

FIGURE 8.5: A momentum-space wave function and its resulting (via a Fourier transform) position-space wave function. The image shows how a non-zero p_0 affects the position-space wave function.

Considering our failure with using only one solution to the Schrödinger equation for the free-particle problem (the lack of localization and normalization), what about a superposition of plane wave solutions which you have explored in Sections 8.3 and 8.4? While these constructions approach a localized solution, there are always copies of the localized solution created. Instead of a sum of individual solutions, consider an integral,

$$\psi(x) = \frac{1}{\sqrt{2\pi\hbar}} \int_{-\infty}^{\infty} \phi(p)\, e^{ipx/\hbar}\, dp\ , \tag{8.8}$$

which is called a Fourier transform. The Fourier transform adds a continuum of plane wave solutions, $e^{ipx/\hbar}$, weighted by a function of momentum, $\phi(p)$. This function of momentum is called the momentum-space wave function since it plays the same role in momentum space as $\psi(x)$ does in position space. The momentum-space wave function, $\phi(p)$, is itself the *inverse* Fourier transform of $\psi(x)$ and is given by:

$$\phi(p) = \frac{1}{\sqrt{2\pi\hbar}} \int_{-\infty}^{\infty} \psi(x)\, e^{-ipx/\hbar}\, dx\ . \tag{8.9}$$

Now, we seek to understand the generic wave function as defined by the Fourier transform in Eq. (8.8) by substituting a reasonable function for $\phi(p)$ and calculating the position-space wave function. Consider a normalized Gaussian distribution in momentum centered on a momentum, p_0, such that

$$\phi(p) = \sqrt{\frac{\alpha}{\sqrt{\pi}}}\, e^{-\alpha^2 (p-p_0)^2/2}\ . \tag{8.10}$$

Note that $|\phi(p)|^2$ goes to $1/e$ of its maximum value when $p = p_0 \pm 1/\alpha$. Therefore $1/\alpha$ tells us something about the spread of the momentum-space wave function. This momentum-space wave function is shown in the bottom panel of the animation. In the animation, $\hbar = 2m = 1$.

To find the position-space wave function, we must use Eq. (8.10) in Eq. (8.8) and evaluate the resulting integral. When we do this Gaussian integral, we get:[4]

$$\psi(x) = \frac{1}{\sqrt{\sqrt{\pi}\alpha\hbar}}\, e^{ip_0x/\hbar - x^2/2\alpha^2\hbar^2} \ . \tag{8.11}$$

Look at the animation to see how the position-space wave function is related to the original momentum-space wave function. The bottom panel shows momentum space and the top panel shows position space. Vary p_0 and α and see what happens. As p_0 gets larger and positive, the momentum-space wave function shifts to the right and is centered on the new value of p_0. The position-space wave function now has bands of color which represent the $e^{ip_0x/\hbar}$ factor in the wave function. As α increases, the momentum-space wave function narrows and the position-space wave function widens (which is a result of the Heisenberg uncertainty principle).

Our packet has almost all of the right features we want in a packet that simulates a *particle*. However, it does not have a time dependence and it does not allow us to shift the initial position, x_0, of the packet to any value we like. We will add these features next.

8.6   THE QUANTUM-MECHANICAL WAVE PACKET SOLUTION

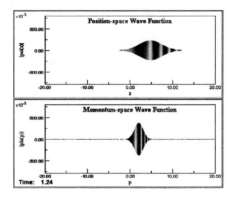

FIGURE 8.6: A momentum-space wave function and its resulting position-space wave function. Both wave functions are allowed to evolve in time according to the Schrödinger equation. Notice the spreading in the position-space wave function and the lack of spread in the momentum-space wave function.

The *time-dependent* position-space wave function can be calculated from the Fourier transform

$$\psi(x,t) = \frac{1}{\sqrt{2\pi\hbar}} \int_{-\infty}^{\infty} \phi(p,t)\, e^{ipx/\hbar}\, dp \ , \tag{8.12}$$

---

[4]Since the momentum-space wave function was normalized, so is the resulting position-space wave function. In general, due to the relationship between $\psi(x)$ and $\phi(p)$ as expressed in Eq. (8.8) and Eq. (8.9), we have that: $\int_{-\infty}^{\infty} |\psi(x)|^2\, dx = \int_{-\infty}^{\infty} |\phi(p)|^2\, dp$, and hence if one is normalized, so is the other.

of the time-dependent momentum-space wave function

$$\phi(p,t) = \sqrt{\frac{\alpha}{\sqrt{\pi}}}\, e^{-\alpha^2(p-p_0)^2/2}\, e^{-ipx_0/\hbar}\, e^{-ip^2t/2m\hbar} \;. \tag{8.13}$$

The added time dependence in Eq. (8.13) is in the term $e^{-ip^2t/2m\hbar}$. Compare this momentum-space wave function to the one we used in Section 8.5. Note that Eq. (8.13) has another additional factor, $e^{-ipx_0/\hbar}$, which sets the center of the Gaussian in position space to x_0. This function is shown in the bottom panel of the animation. When we do the Gaussian integral in Eq. (8.12) using Eq. (8.13) for $\phi(p,t)$, we get:

$$\psi(x,t) = \frac{1}{\sqrt{\sqrt{\pi}\alpha\hbar(1+it/t_0)}}\, e^{ip_0(x-x_0)/\hbar}\, e^{-ip_0^2t/2m\hbar}\, e^{-\frac{(x-x_0-p_0t/m)^2}{2\alpha^2\hbar^2(1+it/t_0)}} \;, \tag{8.14}$$

where we have used the substitution[5] $t_0 = m\hbar\alpha^2$.

Look at the animation to see how this position-space wave function is related to the original momentum-space wave function. The bottom panel shows momentum space and the top panel shows position space. In the animation, $\hbar = 2m = 1$. Vary x_0, p_0, and α and see what happens. We considered the effect of p_0 and α in Section 8.5. As x_0 gets larger and positive, the position-space wave function shifts to the right and is now centered on the new value of x_0. At $t = 0$ the momentum-space wave function has bands of color which represent the $e^{ip_0x/\hbar}$ factor in the wave function. Note the effect of time evolution on the position-space and momentum-space wave functions by pressing the "input values and play" button. The position-space wave function spreads over time but the momentum-space wave function does not, but $\phi(p,t)$ does change phase as a function of time.

We can calculate the probability density, $\psi^*(x,t)\,\psi(x,t)$, as

$$\rho(x,t) = \frac{1}{\alpha\hbar\sqrt{\pi\left(1+(t/t_0)^2\right)}}\, e^{-\frac{(x-x_0-p_0t/m)^2}{\alpha^2\hbar^2(1+(t/t_0)^2)}} \;. \tag{8.15}$$

The maximum height of the probability density is governed by

$$\frac{1}{\alpha\hbar\sqrt{\pi\left(1+(t/t_0)^2\right)}} \;, \tag{8.16}$$

since the exponential's maximum is fixed at 1. The *spread* in x, δx, of the probability density is proportional to the square root of the denominator in the exponential

$$\delta x \propto \alpha\hbar\,\sqrt{1+(t/t_0)^2} \;, \tag{8.17}$$

where $\alpha\hbar$ is the initial ($t=0$) spread of the probability density and is related to the inverse of the spread in momentum. The term in the square root describes the time dependence of the spread in position of the position-space probability density. This time dependence is inversely related to the initial spread in the position-space

---

[5]This substitution is often used in the literature. See, for example, Refs. [16, 17].

probability density. The position of the central peak is where the numerator of the exponential is zero, or where

$$x = x_0 + p_0 t/m . \tag{8.18}$$

Explicitly, from the calculation of expectation values, we see that

$$\langle \hat{x} \rangle = x_0 + p_0 t/m , \qquad \langle \hat{x}^2 \rangle = (x_0 + p_0 t/m)^2 + \frac{\alpha^2 \hbar^2}{2} \left[1 + (t/t_0)^2 \right] , \tag{8.19}$$

$$\langle \hat{p} \rangle = p_0 , \quad \text{and} \quad \langle \hat{p}^2 \rangle = p_0^2 + 1/2\alpha^2 . \tag{8.20}$$

Thus, $\Delta x \equiv \sqrt{\langle \hat{x}^2 \rangle - \langle \hat{x} \rangle^2} = \frac{\alpha \hbar}{\sqrt{2}} \sqrt{1 + (t/t_0)^2}$ and $\Delta p \equiv \sqrt{\langle \hat{p}^2 \rangle - \langle \hat{p} \rangle^2} = \frac{1}{\sqrt{2}\alpha}$ and finally

$$\Delta x \Delta p = \frac{\hbar}{2} \sqrt{1 + (t/t_0)^2} \geq \frac{\hbar}{2} , \tag{8.21}$$

which it must be in order to satisfy the Heisenberg uncertainly principle.[6] Since at $t = 0$, $\Delta x \Delta p = \frac{\hbar}{2}$, the Gaussian wave packet is called a *minimum indeterminacy* wave packet. Finally, the expectation value of the energy for the packet is

$$\langle E \rangle = \langle \hat{T} \rangle = \langle \hat{p}^2 \rangle / 2m = \left[p_0^2 + 1/2\alpha^2 \right] / 2m , \tag{8.22}$$

which tells us that the spread in the momentum distribution of the wave packet also contributes to its average kinetic, $\langle \hat{T} \rangle$, and average total, $\langle E \rangle$, energies.

8.7 EXPLORING FOURIER TRANSFORMS BY MATCHING

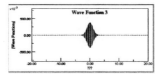

FIGURE 8.7: A (position-space or momentum-space) wave function that must be matched with its corresponding Fourier transformed (momentum-space or position-space) wave function.

Shown are six Gaussian wave functions at $t = 0$ in color-as-phase representation. In the animation, $\hbar = 2m = 1$.

(a) If the wave functions shown are in position space, qualitatively what do the momentum-space wave functions look like?

(b) If the wave functions shown are in momentum space, qualitatively what do the position-space wave functions look like?

Draw your answers making sure to label axes and to represent the phase of the resulting wave function as lines across your wave function. Once you do so, check your answers with the animation provided.

---

[6] By convention, we use the phrase Heisenberg uncertainly principle, but again this concept is best stated as the Heisenberg *indeterminacy* principle since indeterminacy better represents the concept that Heisenberg was trying to describe.

8.8 EXPLORING WAVE PACKETS WITH CLASSICAL ANALOGIES

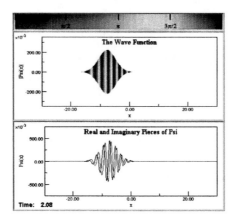

FIGURE 8.8: A quantum-mechanical free particle in color-as-phase representation and in real-and-imaginary component representation.

Shown is a free quantum-mechanical wave packet. The top panel shows the wave functions in color-as-phase representation and the bottom panel shows the wave function in real-and-imaginary component representation. In the animation, $\hbar = 2m = 1$.

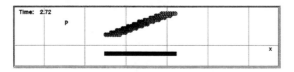

FIGURE 8.9: A classical *packet* of balls of different (constant) velocities.

The second animation, classical *packet* depicts several balls, initially localized, but each row with a different velocity.[7] The gray bar signifies the spread in the packet in the x direction.

Watch both the "classical packet" and "quantum-mechanical packet" animations.

(a) Why does the classical packet spread over time?

(b) Now look at the quantum-mechanical packet again. Why does it spread? Make sure to use quantum-mechanical arguments. Hint: can you come up with an analogy using the classical case shown here?

(c) What is the position-momentum correlation of each packet at $t = 0$? What is the position-momentum correlation of each packet at $t = 2$?

---

[7]This Exploration is based in part on R. W. Robinett's talk, "Quantum Wave Packet Revivals" given at the 128th AAPT National Meeting, Miami Beach, FL, Jan. 24-28, 2004.

PROBLEMS

8.1. This animation shows the real (blue) and imaginary (pink) parts of three wave functions. Assume x is measured in cm and time is measured in seconds.
(a) Write an expression for each of the three wave functions in the form: $e^{i(kx-\omega t)}$.
(b) What is the mass of the particle?
(c) What would the mass of the particle be if time was being shown in ms?

8.2. These animations show four wave functions in *color-as-phase* representation. Assume x is measured in cm and time is measured in seconds.
(a) Rank free particle wave functions 1 through 4 in order of momentum.
(b) Rank free particle wave functions 1 through 4 in order of energy.
Ties in parenthesis please!

8.3. These animations show three wave functions in *color-as-phase* representation. Assume x is measured in cm and time is measured in seconds.

Assume "Wave Function 1" is correct for a particle of mass m. Which of the other functions could be valid free particle wave functions for this same particle?

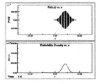

8.4. The animation shows the time development of a free-particle Gaussian wave packet. You may change the parameters: x_0, p_0, and α, then click one of the three links to run the animation. The graph shows the real (blue) and imaginary (pink) parts of the wave function, contributions to the probability density from the real (blue) and imaginary (pink) parts of the wave function, and the total probability density. In the animation, $\hbar = 2m = 1$.

Watch each animation for 1.5 time units before changing values and answering the following questions.
(a) How does the time-evolution of the maximum amplitude and width of the wave packet change as you vary x_0?
(b) How does the time-evolution of the maximum amplitude and width of the wave packet change as you vary p_0?
(c) How does the time-evolution of the maximum amplitude and width of the wave packet change as you vary α?
(d) Explain why this is. Try to be as complete as possible.

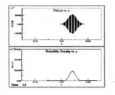

8.5. Shown are five different free quantum-mechanical packets. The top panel shows phase as color and the bottom panel shows the real and imaginary parts of the wave function. In the animation, $\hbar = 2m = 1$.

Rank the packets according to:

(a) their momentum (meaning $\langle \hat{p}_x \rangle$)

(b) their kinetic energy (meaning $\langle \hat{T} \rangle$)

(c) their total energy (meaning $\langle E \rangle = \langle \hat{H} \rangle$)

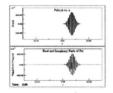

8.6. The animation shows the time development of an accelerating Gaussian wave packet. You may change the parameters: x_0, p_0, α, and a, then click on one of the links to run the animation. The graph shows the real (blue) and imaginary (pink) parts of the wave function, contributions to the probability density from the real (blue) and imaginary (pink) parts of the wave function, and the total probability density. In the animation, $\hbar = 2m = 1$.

Watch each animation for 1.5 time units before changing values and answering the following questions.

(a) How do the time-evolution of the maximum amplitude and width of the wave packet change as you vary x_0?

(b) How do the time-evolution of the maximum amplitude and width of the wave packet change as you vary p_0?

(c) How do the time-evolution of the maximum amplitude and width of the wave packet change as you vary α?

(d) How do the time-evolution of the maximum amplitude and width of the wave packet change as you vary a?

(e) Explain why this is. Try to be as complete as possible.

CHAPTER 9

Scattering in One Dimension

9.1 THE SCATTERING OF CLASSICAL ELECTROMAGNETIC WAVES
9.2 EXPLORING CLASSICAL AND QUANTUM SCATTERING
9.3 THE PROBABILITY CURRENT DENSITY
9.4 PLANE WAVE SCATTERING: POTENTIAL ENERGY STEPS
9.5 EXPLORING THE ADDITION OF TWO PLANE WAVES
9.6 PLANE WAVE SCATTERING: FINITE BARRIERS AND WELLS
9.7 EXPLORING SCATTERING AND BARRIER HEIGHT
9.8 EXPLORING SCATTERING AND BARRIER WIDTH
9.9 EXPLORING WAVE PACKET SCATTERING

INTRODUCTION

We now consider another one-dimensional problem, the scattering problem. In doing so we need to consider scattering-type solutions and what they mean. For standard scattering situations, the wave functions we use are usually those valid for regions of constant potential energy such as complex exponentials (plane waves) when $E > V_0$ and real exponentials when $E < V_0$.[1]

9.1 THE SCATTERING OF CLASSICAL ELECTROMAGNETIC WAVES

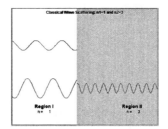

FIGURE 9.1: A classical electromagnetic wave incident on an *increase* in index of refraction. The incident, transmitted, and reflected waves are all shown.

A classical electromagnetic wave at an interface where there is a change of index of refraction is similar to, but not exactly the same as, a quantum-mechanical plane wave incident on a change in potential energy.

---

[1]There is one other possibility that is not often considered. If $E = V_0$, the solution to the Schrödinger equation yields a linear solution.

Change the index of refraction in the region on the right, Region II, and see what happens to the incident right-moving, transmitted right-moving, and reflected left-moving waves. Clicking in the check box enables you to see the sum of the incident and the reflected waves in Region I. Note that the reflected wave travels in the opposite direction from the incident wave and is 180° out of phase with the incident wave at the interface.

The amplitude of the reflected wave becomes larger as n_2 increases. When you add the incident wave to the phase-shifted reflected wave as the two amplitudes approach each other with increasing n_2, the resultant wave in Region I will resemble a standing wave. This idea is important for the visualization of quantum-mechanical barrier problems. When an electromagnetic wave travels from a region of smaller n to a region of larger n, its wavelength and speed decrease. In addition, the amplitude of the transmitted electromagnetic wave becomes smaller as n increases.

Now change the index of refraction in the region on the left, Region I, and see what happens to the incident right-moving, transmitted right-moving, and reflected left-moving waves. The amplitude of the reflected wave becomes larger as n_1 increases. When you add the incident wave to the reflected wave as n_1 increases, the resultant wave in Region I will again resemble a standing wave. However, when an electromagnetic wave travels from a region of larger n to a region of smaller n, its wavelength and speed increase. In addition, the amplitude of the transmitted electromagnetic wave becomes larger as n increases.

The reflected wave's amplitude (and hence energy) is also related to the energy stored in the region (the dielectric).

9.2   EXPLORING CLASSICAL AND QUANTUM SCATTERING

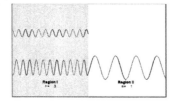

FIGURE 9.2: A classical electromagnetic wave incident on an *decrease* in index of refraction. The incident, transmitted, reflected, and the sum of the incident and reflected waves are all shown.

This Exploration stresses the similarities and differences between a classical electromagnetic wave incident on a change (an increase or decrease) of index of refraction and a quantum-mechanical plane wave incident on a change (an increase or decrease) in potential energy. Use the check boxes to switch between classical and quantum-mechanical waves to see the result of the sum of the incident and reflected electromagnetic waves in Region I.

Answer the following questions for both the case of $n_1 < n_2$ and $n_1 > n_2$ and $V_1 < V_2$ and $V_1 > V_2$.

(a) What is the phase of the reflected wave relative to the incident wave in the classical and quantum-mechanical cases?

(b) What happens to the amplitude of the wave in Region II for the classical and quantum-mechanical cases?

(c) As $n_2 >> n_1$ and as $V_2 >> V_1$, what does the superposition of the incident and reflected waves look like in the classical and quantum-mechanical cases?

(d) What happens to the wavelength and speed of the electromagnetic wave in Region II as compared to Region I? What happens to the *curviness* and the momentum of the quantum-mechanical plane wave in Region II as compared to Region I?

Note that the quantum-mechanical case is considered in detail beginning in Section 9.4.

9.3 THE PROBABILITY CURRENT DENSITY

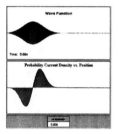

FIGURE 9.3: A Gaussian wave packet in an infinite square well is shown using color-as-phase representation. Accompanying the wave function is the quantum-mechanical probability current density and the calculation of the probability current.

The animation depicts a Gaussian wave packet in an infinite square well. The packet has no initial momentum, but after its left edge *hits* the infinite wall, the packet begins to move to the right. The animation also shows the probability current density and the total probability current. In the animation, $\hbar = 2m = 1$.

What is the probability current density? We can construct a probability relationship from the one-dimensional Schrödinger equation which, for real $V(x)$, yields[2]

$$\partial J_x / \partial x + \partial \rho / \partial t = 0 \ . \tag{9.3}$$

This relationship takes the same form as the differential statement of charge-current conservation from electromagnetism. Here, however $\rho(x,t)$ and $J_x(x,t)$ are not the

---

[2]Consider the construction: $\psi^*(x,t)\, SE - SE^*\, \psi(x,t)$ where the complex conjugate of the Schrödinger equation, SE^*, with $V(x)$ real, is

$$\frac{-\hbar^2}{2m}\, \frac{\partial^2}{\partial x^2}\, \psi^*(x,t) + V(x)\psi^*(x,t) = i\hbar \frac{\partial}{\partial t}\, \psi^*(x,t) \ . \tag{9.1}$$

This ultimately, after a fair amount of algebra, yields:

$$\frac{\partial}{\partial x}\left[\frac{\hbar}{2mi}\left\{\psi^*(x,t)\,[\frac{\partial}{\partial x}\,\psi(x,t)] - [\frac{\partial}{\partial x}\psi^*(x,t)]\,\psi(x,t)\right\}\right] + \frac{\partial}{\partial t}\,[\psi^*(x,t)\,\psi(x,t)] = 0 \ , \tag{9.2}$$

again as long as $V(x)$ is real.

electric charge and electric current densities of electromagnetism, they are instead

$$\rho(x,t) = \psi^*(x,t)\,\psi(x,t)\,, \tag{9.4}$$

and

$$J_x(x,t) = \frac{\hbar}{2mi}\left\{\psi^*(x,t)\left[\partial\psi(x,t)/\partial x\right] - \left[\partial\psi^*(x,t)/\partial x\right]\psi(x,t)\right\}, \tag{9.5}$$

which are the probability density and the probability current density in one dimension, respectively.[3]

As in Section 7.6, consider an arbitrary complex wave function that we choose to write as $\psi(x,t) = A(x,t)\,e^{i\theta(x,t)}$. Consider the probability density and the probability current density in this representation:

$$\rho(x,t) = \psi^*(x,t)\,\psi(x,t) = A^2(x,t) \quad \text{and} \quad J_x = \frac{\hbar}{m}A^2(x,t)\left[\partial\theta(x,t)/\partial x\right]. \tag{9.6}$$

This tells us that the probability density is insensitive to the phase of the wave function, but the probability current density is sensitive to *spatial changes* in phase. Given that the phase is represented by color in this animation, spatial variations in the *color* of $\psi(x,t)$ signify a non-zero probability current density. Restart the animation and watch how the probability current density evolves with time.

The insensitivity of the probability density to phase may suggest that the phase of the wave function is unimportant (even though the probability current *does* depend on the phase). When we consider wave functions that are superpositions of two or more states, we will find that *relative* phase does matter.

9.4   PLANE WAVE SCATTERING: POTENTIAL ENERGY STEPS

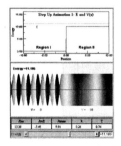

FIGURE 9.4: A quantum-mechanical plane wave shown in color-as-phase representation incident on a *step up* in potential energy. The energy of the wave function is greater than that of the step height.

We begin our study of quantum-mechanical scattering by looking at a general case where we have two regions (Regions I and II) and a change in the potential energy at the interface between the two regions. Typically we choose $V = 0$ in

---

[3]In three-dimensions, Eqs. (9.3), (9.4), and (9.5) can be generalized to $\nabla \cdot \mathbf{J} + \partial\rho/\partial t = 0$, $\rho(\mathbf{r},t) = \psi^*(\mathbf{r},t)\,\psi(\mathbf{r},t)$, and $\mathbf{J}(\mathbf{r},t) = \frac{\hbar}{2mi}\left\{\psi^*(\mathbf{r},t)\left[\nabla\psi(\mathbf{r},t)\right] - \left[\nabla\psi^*(\mathbf{r},t)\right]\psi(\mathbf{r},t)\right\}$.

Region I and $V = V_0$ in Region II, and the change occurs at $x = 0$. This situation is shown in "Step Up Animation 1" and "Step Up Animation 2." The wave functions when $E > V_0$ are

$$\psi_{\text{inc}} = Ae^{i(k_1 x - \omega t)} , \quad \psi_{\text{refl}} = Be^{-i(k_1 x + \omega t)} , \quad \text{and} \quad \psi_{\text{trans}} = Ce^{i(k_2 x - \omega t)} , \quad (9.7)$$

where $k_1^2 = 2mE/\hbar^2$ and $k_2^2 = 2m(E - V_0)/\hbar^2$, while the momenta are just: $p_{\text{inc}} = \hbar k_1$, $p_{\text{refl}} = -\hbar k_1$, and $p_{\text{trans}} = \hbar k_2$. In addition, we note that the ω's are the same. Why? The ω's are related to the *total energy* of the plane wave. The *total energy* must be the same in each region (the entire wave function represents one energy eigenstate of the Schrödinger equation) even when the k's are different in each region. Therefore, since $E = \hbar\omega$, the ω's must be the same. In addition, if the ω's were different, the solution to the boundary-value problem we will shortly consider would not necessarily be valid for all times.

To construct transmission and reflection coefficients, we must first calculate J_x for the incident, reflected, and transmitted waves from the equation[4]

$$J_x = \frac{\hbar}{2mi} \left[\psi^* \frac{d\psi}{dx} - \frac{d\psi^*}{dx} \psi \right] . \quad (9.11)$$

We find that

$$J_{\text{inc}} = \frac{\hbar k_1}{m} |A|^2 , \quad J_{\text{refl}} = -\frac{\hbar k_1}{m} |B|^2 , \quad \text{and} \quad J_{\text{trans}} = \frac{\hbar k_2}{m} |C|^2 . \quad (9.12)$$

For $E > V$ we may use the above *parts of the wave function*, but we must match the wave function parts at the change in potential energy (the boundary at $x = 0$) since the parts in Eq. (9.7) represent just one wave function.[5] We require that $\psi_{\text{I}}(0) = \psi_{\text{II}}(0)$ and that $\psi_{\text{I}}'(0) = \psi_{\text{II}}'(0)$. We can use time-independent wave functions (since the time dependence for all wave function parts is identical, it cancels):

$$\psi_{\text{inc}} = Ae^{ik_1 x} , \quad \psi_{\text{refl}} = Be^{-ik_1 x} , \quad \text{and} \quad \psi_{\text{trans}} = Ce^{ik_2 x} , \quad (9.13)$$

---

[4]You may be wondering why we care about the probability current density. The conservation of probability in its integral form in one dimension is

$$\frac{dP}{dt} + \int_{-\infty}^{\infty} \frac{dJ_x}{dx} \, dx = 0 , \quad (9.8)$$

and for a scattering process, the total probability of finding the particle over all space does not change with time, or that $dP/dt = 0$. Therefore: $\int_{-\infty}^{\infty} \frac{dJ_x}{dx} \, dx = 0$ or that $J_x(\infty) = J_x(-\infty)$. We can identify $J_x(\infty) \equiv J_{\text{trans}}$ and $J_x(-\infty) \equiv J_{\text{inc}} + J_{\text{refl}}$. We find that $J_{\text{trans}} = J_{\text{inc}} + J_{\text{refl}}$ and therefore $J_{\text{inc}} = J_{\text{trans}} - J_{\text{refl}}$, which leads directly to

$$1 = J_{\text{trans}}/J_{\text{inc}} - J_{\text{refl}}/J_{\text{inc}} . \quad (9.9)$$

The sign in front of the J_{refl} term is due to the fact that this expression is always *negative*, therefore $-J_{\text{refl}}$ will be *positive*. We incorporate this sign in the definition of the reflection and transmission coefficients as

$$T = J_{\text{trans}}/J_{\text{inc}} \equiv |J_{\text{trans}}/J_{\text{inc}}| \quad \text{and} \quad R = -J_{\text{refl}}/J_{\text{inc}} \equiv |J_{\text{refl}}/J_{\text{inc}}| . \quad (9.10)$$

[5]Also look for the $E = V_0$ case in the animations. The solution in Region II should be a straight line as discussed.

where $\psi_I = \psi_{inc} + \psi_{refl}$. When we match $\psi_I(0) = \psi_{II}(0)$, this gives $A + B = C$. When we match $\psi'_I(0) = \psi'_{II}(0)$, we get $k_1 A - k_1 B = k_2 C$. From these conditions we solve for the fractions $|B/A|^2$ and $|C/A|^2$

$$|B/A|^2 = (1 - k_2/k_1)^2/(1 + k_2/k_1)^2 , \quad \text{and} \quad |C/A|^2 = 4/(1 + k_2/k_1)^2 , \quad (9.14)$$

since they occur in the the current densities, Eq. (9.12), which are necessary in order to calculate the transmission and reflection coefficients from Eq. (9.10). We therefore find that

$$R = (1 - k_2/k_1)^2/(1 + k_2/k_1)^2 \qquad (9.15)$$

and

$$T = \left(\frac{k_2}{k_1}\right) \frac{4}{(1 + k_2/k_1)^2} . \qquad (9.16)$$

Recall $k_1^2 = 2mE/\hbar^2$ and $k_2^2 = 2m(E - V_0)/\hbar^2$. Therefore, $k_2^2/k_1^2 = (E - V_0)/E = 1 - V_0/E$. As E increases, $k_2 \to k_1$, and therefore $T \to 1$, as expected. With a bit of algebra, you can also convince yourself that $T + R = 1$ by adding Eqs. (9.15) and (9.16) together.

If we had the opposite circumstance, as shown in "Step Down Animation 1" and "Step Down Animation 2," with an incident plane wave subject to a potential energy V_0 and then at $x > 0$ experiencing no potential energy, this would amount to $k_2 \to k_1$ and $k_1 \to k_2$. Multiplying this result by the square of the ratio of the k's gets us back to the original T and R in Eqs. (9.15) and (9.16). Therefore the transmission and reflection coefficients for these two different problems are actually the same.

Look at both "Step Up Animation 1" and "Step Up Animation 2." "Step Up Animation 1" shows you the quantum-mechanical plane wave in both regions and the top panel shows the energy diagram. Change the energy of the incident plane wave by moving the slider and note the effect on the plane wave and also the transmission and reflection coefficients. "Step Up Animation 2" shows the same situation but with a graph of the transmission and reflection coefficients as a function of energy. Click-drag the plane wave and drag up to increase the energy or drag down to decrease the energy. If you do this slowly enough, you can create a very nice graph of the transmission and reflection coefficients versus E for a fixed V_0.

For $E < V_0$, we can use the time-independent wave functions

$$\psi_{inc} = Ae^{ik_1 x} , \quad \psi_{refl} = Be^{-ik_1 x} , \quad \text{and} \quad \psi_{trans} = Ce^{-\kappa_2 x} , \qquad (9.17)$$

where $k_1^2 = 2mE/\hbar^2$ and $\kappa_2^2 = 2m(|E - V_0|)/\hbar^2$. We find for this case

$$J_{inc} = \frac{\hbar k_1}{m}|A|^2 , \quad J_{refl} = -\frac{\hbar k_1}{m}|B|^2 , \quad \text{and} \quad J_{trans} = 0 . \qquad (9.18)$$

Now we must again match the parts of the wave function. We require that $\psi_I(0) = \psi_{II}(0)$ and that $\psi'_I(0) = \psi'_{II}(0)$. When we match $\psi_I(0) = \psi_{II}(0)$, this gives

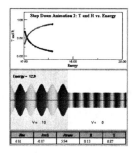

FIGURE 9.5: A quantum-mechanical plane wave shown in color-as-phase representation incident on a *step down* in potential energy. The energy of the wave is greater than that of the step height. A graph of the transmission and reflection coefficients is shown as a function of wave function energy.

$A + B = C$. When we match $\psi_I'(0) = \psi_{II}'(0)$, we get $ik_1 A - ik_1 B = -\kappa_2 C$. We eventually arrive at the relationship

$$|B/A|^2 = \frac{(k_1 + i\kappa_2)}{(k_1 - i\kappa_2)} \frac{(k_1 - i\kappa_2)}{(k_1 + i\kappa_2)} = 1 . \tag{9.19}$$

Note in this case, $R = |B/A|^2 = 1$. From $R = 1$ we also expect $T = 0$, which is also clear from the fact that $J_{\text{trans}} = 0$. In addition, if $E = V_0$, we find that $T = 0$. These results agree with the classical scattering case *except* that there is a finite probability of finding the particle in the *classically-forbidden* region.

9.5   EXPLORING THE ADDITION OF TWO PLANE WAVES

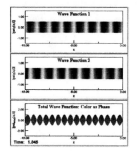

FIGURE 9.6: Two quantum-mechanical plane waves shown in color-as-phase representation. The sum of the two plane waves is shown in the bottom graph.

During the quantum-mechanical scattering of plane waves, we must consider the superposition of the right-moving incident wave and the left-moving reflected wave results in Region I. In this Exploration we investigate how two time-dependent plane waves can be added together to resemble a scattering situation. In the animation, $\hbar = 2m = 1$.

(a) With the default settings, explain why the arguments of the cosines and sines are of the form (5*x-25*t) and (-5*x-25*t). In other words, what does the

± 5 signify and what does the 25 signify? Remember that $\hbar = 2m = 1$ in this animation.

(b) With the default settings, describe the sum of the two plane waves. Look at the real and imaginary parts of the wave functions to verify your conjecture.

(c) With the default settings, if this animation were showing the wave function in Region I of a scattering problem, what would T and R be?

(d) Now set re2 = -0.9*cos(-5*x-25*t) and im2 = -0.9*sin(-5*x-25*t) for Wave Function 2. What results? Now change the number multiplying Wave Function 2 to -0.8, -0.7, -0.6, -0.5, -0.4, -0.3, -0.2, -0.1, and finally 0. What happens to the resulting wave as you change this amplitude? If Wave Function 1 represented ψ_{inc} and Wave Function 2 represented ψ_{refl}, what would happen to the transmission coefficient as you change the amplitude of Wave Function 2?

9.6 PLANE WAVE SCATTERING: FINITE BARRIERS AND WELLS

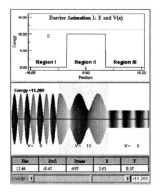

FIGURE 9.7: A quantum-mechanical plane wave shown in color-as-phase representation incident on a finite barrier in potential energy. The energy of the wave is greater than that of the barrier height.

We now consider a barrier that exists only from $-a < x < a$ with a height V_0. This situation is shown in "Barrier Animation 1" and "Barrier Animation 2." This problem proceeds much like the previous cases, except there are now *three* regions of interest.

For $E > V_0$, we consider the wave functions in each region:

$$\psi_{\text{I inc}} = Ae^{i(k_1 x - \omega_1 t)} , \qquad \psi_{\text{I refl}} = Be^{-i(k_1 x + \omega_1 t)} , \qquad (9.20)$$

$$\psi_{\text{II R}} = Ce^{i(k_2 x - \omega_2 t)} , \qquad \psi_{\text{II L}} = De^{-i(k_2 x + \omega_2 t)} , \qquad (9.21)$$

and

$$\psi_{\text{III trans}} = Fe^{i(k_1 x - \omega_1 t)} , \qquad (9.22)$$

where $k_1^2 = 2mE/\hbar^2$ and $k_2^2 = 2m(E - V_0)/\hbar^2$. We calculate J_x for the incident, reflected, and transmitted waves:

$$J_{\text{inc}} = \frac{\hbar k_1}{m}|A|^2 , \quad J_{\text{refl}} = -\frac{\hbar k_1}{m}|B|^2 , \quad \text{and} \quad J_{\text{trans}} = \frac{\hbar k_1}{m}|F|^2 , \qquad (9.23)$$

and since $k_1 = k_3$, we find that the transmission and reflection coefficients reduce to just $T = |F/A|^2$ and $R = |B/A|^2$.

To find these ratios, we must again match the pieces of the wave function at the boundaries of the regions ($\psi_{\mathrm{II}} = \psi_{\mathrm{II\,R}} + \psi_{\mathrm{II\,L}}$). We require that $\psi_{\mathrm{I}}(-a) = \psi_{\mathrm{II}}(-a)$, $\psi'_{\mathrm{I}}(-a) = \psi'_{\mathrm{II}}(-a)$, $\psi_{\mathrm{II}}(a) = \psi_{\mathrm{III}}(a)$, and $\psi'_{\mathrm{II}}(a) = \psi'_{\mathrm{III}}(a)$. We get the following result after much algebra:

$$(A/F) = \left[\cos(2k_2 a) - i\left(\frac{k_1^2 + k_2^2}{2k_1 k_2}\right)\sin(2k_2 a)\right] \qquad (9.24)$$

and therefore for $1/T$ we have

$$1/T = |(A/F)|^2 = \left[1 + \left(\frac{k_1^2 - k_2^2}{2k_1 k_2}\right)^2 \sin^2(2k_2 a)\right]. \qquad (9.25)$$

We note that $T \to 1$ when E is large enough ($E \gg V_0$ and therefore $k_2 \to k_1$) and also when the sine term is zero (when $2k_2 a = n\pi$ or that $k_2 = n\pi/2a = 2\pi/\lambda_2$). Now, since $k_1^2 = 2mE/\hbar^2$ and $k_2^2 = 2m(E - V_0)/\hbar^2$, we can rewrite $1/T$ in terms of E and V_0 as

$$1/T = |(A/F)|^2 = \left[1 + \left(\frac{V_0^2}{4E(E - V_0)}\right)\sin^2(2k_2 a)\right]. \qquad (9.26)$$

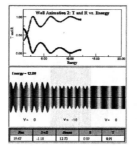

FIGURE 9.8: A quantum-mechanical plane wave shown in color-as-phase representation incident on a finite potential energy well. A graph of the transmission and reflection coefficients is shown as a function of wave function energy.

Now, for $E < V_0$ we have a similar situation.[6] We consider the barrier that exists again only from $-a < x < a$ with a height V_0. This problem proceeds much like the $E > V_0$ case we have considered before except in the second region we make the substitution $ik_2 \to \kappa_2$ or that $k_2 \to -i\kappa_2$ and therefore

$$\psi_{\mathrm{II\,R}} = Ce^{-\kappa_2 x - i\omega_2 t} \qquad \text{and} \qquad \psi_{\mathrm{II\,L}} = De^{\kappa_2 x - i\omega_2 t}. \qquad (9.27)$$

Using the substitution $k_2 \to -i\kappa_2$, the transmission coefficient becomes:

$$1/T = |(A/F)|^2 = \left[1 + \left(\frac{k_1^2 + \kappa_2^2}{2k_1 \kappa_2}\right)^2 \sinh^2(2\kappa_2 a)\right], \qquad (9.28)$$

---

[6]Again, there is one more possibility, $E = V_0$. Look for this in the animations.

since $\sin(iz) = i\sinh(z)$. We can rewrite $1/T$ in terms of E and V_0 using $k_1^2 = 2mE/\hbar^2$ and $\kappa_2^2 = 2m(V_0 - E)/\hbar^2$ as

$$1/T = |(A/F)|^2 = \left[1 + \left(\frac{V_0^2}{4E(V_0 - E)}\right)\sinh^2(2\kappa_2 a)\right]. \tag{9.29}$$

Notice that Eq. (9.29) implies for $E < V_0$, $T < 1$, but T is not necessarily zero.

Scattering from a well, for $E > 0$ but with $V_0 < 0$ or $V_0 = -|V_0|$, we have a situation which is shown in "Well Animation 1" and "Well Animation 2." We consider the well that exists only from $-a < x < a$. We may use the previous answer for the barrier with $E > V_0$:

$$1/T = |(A/F)|^2 = \left[1 + \left(\frac{k_1^2 - k_2^2}{2k_1 k_2}\right)^2 \sin^2(2k_2 a)\right], \tag{9.30}$$

and in terms of E and V_0

$$1/T = |(A/F)|^2 = \left[1 + \left(\frac{V_0^2}{4E(E + |V_0|)}\right)\sin^2(2k_2 a)\right], \tag{9.31}$$

where $k_2^2 = 2m(E + |V_0|)/\hbar^2$. When does $T \to 1$? Look at the second term in Eq. (9.31). When that term goes to zero, $T \to 1$. This happens when $E >> V_0$ and in one other set of occurrences. If we now consider the sine function, it is zero when $n\pi = 2k_2 a$ and therefore $k_2 = n\pi/2a = 2\pi/\lambda_2$, where $2a$ is the width of the well. Now given the relationship $k_2 = n\pi/2a$ that we can insert into $E + |V_0| = \hbar^2 k_2^2/2m$, we get $E + |V_0| = \hbar^2 n^2 \pi^2/2m(2a)^2 = n^2 \tilde{E}_1$, where $\tilde{E}_1$ is the energy of an infinite square well of width $2a$. This effect is called the Ramsauer effect. This effect describes the scattering of low-energy electrons from atoms.

9.7 EXPLORING SCATTERING AND BARRIER HEIGHT

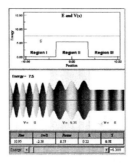

FIGURE 9.9: A quantum-mechanical plane wave shown in color-as-phase representation incident on a finite potential energy barrier. The energy of the wave is greater than that of the barrier height.

Shown in the animation is a plane wave incident on a potential energy barrier. Using the slider you can change the barrier height, V_0, to either above or below the energy of the plane wave which is set at $E = 7.5$. Shown in the table are the probability current densities and the transmission and reflection coefficients. In the animation, $\hbar = 2m = 1$.

(a) For $E < V_0$, what happens as V_0 increases? What do you notice about how the wave function in the barrier behaves as a function of barrier height? What do you notice about how the transmission coefficient behaves?

(b) For $E > V_0$, what happens as V_0 increases? What do you notice about how the wave function in the barrier behaves? What do you notice about how the reflection coefficient behaves as a function of barrier height?

(c) For $E = V_0$, what do you notice about how the transmission coefficient behaves? Does this agree with what you expect from the formula for the transmission coefficient? What do you notice about how the wave function in the barrier behaves?

9.8 EXPLORING SCATTERING AND BARRIER WIDTH

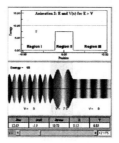

FIGURE 9.10: A quantum-mechanical plane wave shown in color-as-phase representation incident on a finite barrier in potential energy. The energy of the wave is greater than that of the barrier height. A graph of the transmission and reflection coefficients are shown as a function of wave function energy.

Shown in the animation is a plane wave incident on a barrier in potential energy. Using the slider you can change the barrier width. By selecting either "Animation 1" or "Animation 2" you can view the animation with the energy of the plane wave below or above the barrier height, $V_0 = 10$. Shown in the table are the probability current densities and the transmission and reflection coefficients. In the animation, $\hbar = 2m = 1$.

(a) For $E < V_0$, what happens as the barrier width decreases? What do you notice about how the wave function in the barrier behaves as a function of barrier width? What do you notice about how the transmission coefficient behaves?

(b) For $E > V_0$, what happens as the barrier width decreases? What do you notice about how the wave function in the barrier behaves? What do you notice about how the transmission coefficient behaves as a function of barrier width?

9.9   EXPLORING WAVE PACKET SCATTERING

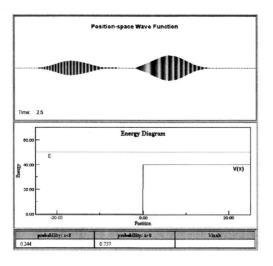

FIGURE 9.11: A free Gaussian wave packet shown in position and momentum space using color-as-phase representation incident scattering from a *step up* in potential energy.

This Exploration shows the *same* initial Gaussian wave packet incident on either a step up in potential energy, a potential energy barrier, or a potential energy well. The animations successively increase the step/barrier height and well depth. Shown in the table are the probabilities that the wave packet can be found in the various regions.[7] In the animation, $\hbar = 2m = 1$.

(a) For the step up animations, what happens to the packet as E increases? What do you notice about how the wave function behaves in the two regions as a function of E? Make a reasonable estimate for how the transmission coefficient behaves as a function of E.

(b) For the barrier animations, what happens to the packet as E increases? What do you notice about how the wave function behaves in the three regions as a function of E? Make a reasonable estimate for how the transmission coefficient behaves as a function of E. What happens to the part of the wave function in the barrier?

(c) For the well animations, what happens to the packet as E increases? What do you notice about how the wave function behaves in the three regions as a function of E? Make a reasonable estimate for how the transmission coefficient behaves as a function of E. What happens to the part of the wave function in the well?

---

[7]The problem of wave packets incident on a step up in potential energy is discussed by R. Shankar in *Principles of Quantum Mechanics*, Plenum Press, 1994.

PROBLEMS

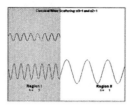

9.1. A classical electromagnetic wave (infrared light) is traveling to the right through two media (I and II) of different indices of refraction n_1 and n_2, respectively (position is given in microns, μm). As the light encounters a change in index of refraction, part of it is transmitted through the surface and part is reflected backwards. Shown in red is the left-moving part of the EM wave (the reflected wave) and shown in blue is the right-moving part of the EM wave (the incident wave in Region I and the transmitted wave in Region II).

Vary n_2 from 1 to 5, keeping $n_1 = 1$ and vary n_1 from 1 to 5, keeping $n_2 = 1$ by selecting one of the links. Answer the following questions for each situation.

(a) How does the reflected wave in Region I compare to the incident wave in Region I? Be as explicit as possible.

(b) As n_2 (n_1) increases, what happens to the reflected wave compared to the transmitted wave?

(c) Consider the case when $n_2 = 5$ ($n_1 = 5$). Looking at the waves in Region I, if they were combined, what would the combination look like?

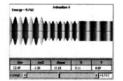

9.2. Drag the slider to change the energy of the incident particle and have the table fill with values. Press the play button to evolve the wave function in time. In the animations, $\hbar = 2m = 1$.

Watch all six animations. Describe each of the potential energy functions that the incident plane wave is experiencing.

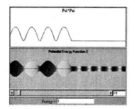

9.3. Drag the slider to change the energy of the incident wave. In the animation, $\hbar = 2m = 1$. Watch all four animations. Describe each of the potential energy functions that the incident plane wave is experiencing. Explain why you answered in this way.

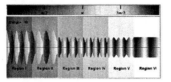

9.4. Shown is a plane wave incident on several different regions. Each region has a constant potential energy associated with it. In the animation, $\hbar = 2m = 1$. You may switch between phase-as-color and real and imaginary representations by selecting a link above. Rank the regions by:

(a) the kinetic energy of the plane wave in that region.

(b) the potential energy of the plane wave in that region.

(c) the energy of the plane wave in that region.

Ties in ( ) please.

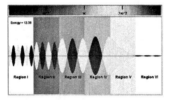

9.5. Shown is a plane wave incident on several different regions. Each region has a constant potential energy associated with it. In the animation, $\hbar = 2m = 1$. You may click-drag within the applet (up or down) to change the energy of the incident plane wave. Regions I and VI have a $V = 0$. Determine the potential energy height, V_i, associated with each of the remaining regions.

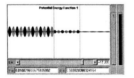

9.6. Drag the slider to change the energy of the incident wave. In the animation, $\hbar = 2m = 1$.

(a) Describe the potential energy function that the incident plane wave experiences for each animation.

(b) How does the potential energy function affect the transmission and reflection coefficients for each animation?

C H A P T E R 10

The Infinite Square Well

10.1 CLASSICAL PARTICLES AND WAVE PACKETS IN AN INFINITE WELL
10.2 THE QUANTUM-MECHANICAL INFINITE SQUARE WELL
10.3 EXPLORING CHANGING WELL WIDTH
10.4 TIME EVOLUTION
10.5 CLASSICAL AND QUANTUM-MECHANICAL PROBABILITIES
10.6 TWO-STATE SUPERPOSITIONS
10.7 WAVE PACKET DYNAMICS
10.8 EXPLORING WAVE PACKET REVIVALS WITH CLASSICAL ANALOGIES

INTRODUCTION

The infinite square well is the prototype bound-state quantum-mechanical problem. Despite this being a *standard* problem, there are still many interesting subtleties of this model for the student to discover. Our understanding of this problem, whether it be in regards to energy eigenstates in position or momentum space, time evolution, or the dynamics of wave packets, will be useful for the study of more realistic problems.

10.1 CLASSICAL PARTICLES AND WAVE PACKETS IN AN INFINITE WELL

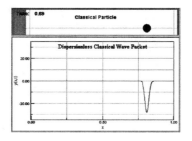

FIGURE 10.1: A classical particle and a *dispersionless* classical wave packet constrained to move in an infinite potential well. The particle and the wave packet have already encountered the infinite wall on the right and have elastically bounced/reflected.

We are going to study the one-dimensional infinite square well problem and use it to illustrate many quantum-mechanical principles. First consider a classical particle confined to a box of length L by an infinite potential well:

$$V = \infty \quad x \le 0, \qquad V = 0 \quad 0 < x < L, \qquad V = \infty \quad x \ge L. \tag{10.1}$$

Classically, we would try to use the relationship between the force and the potential energy: $F_x = -dV/dx$ and then get $x(t)$ and $v(t)$ from $F_x = ma_x$. However, because Eq. (10.1) is such a *badly-behaved* potential energy function, this is impossible to do. Thankfully, determining the trajectory of the particle is relatively straightforward. The classical particle moves freely at a constant velocity (since its potential energy is constant) until it elastically collides with a boundary wall, in which case its velocity changes sign and the particle continues on its way with the exact same speed moving in the opposite direction. This motion is shown in the animation. Think about how you would describe the particle's position and momentum as a function of time.

Another similar classical situation is that of a wave packet on an idealized string of length L with its ends fixed. Wave packet motion of this type is also shown in the animation.[1] Note how the wave packet moves and that as it encounters the end of the string it reflects.

10.2   THE QUANTUM-MECHANICAL INFINITE SQUARE WELL

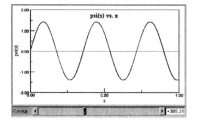

FIGURE 10.2: A solution of the Schrödinger equation satisfying the infinite square well boundary condition at $x = 0$, but not the boundary condition at $x = L$.

In the infinite square well potential, a particle is confined to a box of length L by two infinitely high potential energy barriers:

$$V(x) = \begin{cases} \infty & \text{for} \quad x \le 0 \\ 0 & \text{for} \quad 0 < x < L \\ \infty & \text{for} \quad x \ge L \end{cases} . \tag{10.2}$$

We begin with the time-independent Schrödinger equation in one dimension for $0 < x < L$:

$$-\frac{\hbar^2}{2m}\frac{d^2}{dx^2}\,\psi(x) = E\,\psi(x). \tag{10.3}$$

The solution to this differential equation is a combination of sines and cosines (or complex exponentials; here, because of the boundary conditions, we choose sines

---

[1]This wave is dispersionless. Waves with dispersion are covered in Section 5.11.

and cosines). The general solution is

$$\psi(x) = A\sin(kx) + B\cos(kx) \, , \qquad (10.4)$$

where $k^2 = 2mE/\hbar^2$.

We still need to satisfy the boundary conditions, and thereby determine A, B, and k (and therefore E). For the wall on the left we need $\psi(x) = 0$, for $x \leq 0$ and for the wall on the right we need $\psi(x) = 0$ for $x \geq L$, which means that

$$\psi(0) = 0 \quad \text{and} \quad \psi(L) = 0 \, . \qquad (10.5)$$

The cosine part of the general solution does not vanish at the $x = 0$ boundary since, $\cos(0) \neq 0$, and therefore we must have $B = 0$. We are left with $\psi(x) = A\sin(kx)$ and determining both A and k. This situation is shown in the animation. The boundary condition at $x = 0$ is already solved and you can vary the energy to see the effect on the wave function (this is the shooting method). In the animation, $\hbar = 2m = 1$. For what values of the energy (and therefore k) is the boundary condition at $x = L$ satisfied?

At the boundary $x = L$ we must have a k such that $\psi(L) = A\sin(kL) = 0$. Satisfying this boundary condition can be accomplished by requiring that $kL = n\pi$, where $n = 0, \pm 1, \pm 2, \pm 3, \ldots$. We can eliminate the negative values of n as these wave functions are different from the positive n values by just an overall phase factor of -1. The $n = 0$ value must be considered more carefully. If we allow $n = 0$, this yields $k = 0$ and amounts to a zero-curvature solution to the time-independent Schrödinger equation. A zero-curvature solution is of the form $Ax + B$ and cannot satisfy the boundary conditions and be non-zero in the well. Therefore, $k = 0$ is not a valid possibility[2] and we have that:

$$\psi_n(x) = A\sin(n\pi x/L) \quad \text{for } 0 < x < L \quad \text{with} \quad n = 1, 2, 3, \ldots, \qquad (10.6)$$

and zero otherwise. Since $k = \sqrt{2mE/\hbar^2}$, the energy becomes: $E_n = n^2\pi^2\hbar^2/2mL^2$.

But what about A? The wave function must be normalized to satisfy Born's probabilistic interpretation, so

$$\int_{-\infty}^{\infty} \psi_n^*(x)\,\psi_n(x)\,dx = 1 \, . \qquad (10.7)$$

Since $\psi_n(x)$ is a real function and most of the spatial integral vanishes because the wave function is zero everywhere except between 0 and L, we are simply left with calculating

$$A^2 \int_{-\infty}^{\infty} \sin^2(n\pi x/L)\,dx \quad \rightarrow \quad A^2 \int_0^L \sin^2(n\pi x/L)\,dx = A^2\frac{L}{2} = 1 \, , \qquad (10.8)$$

---

[2]Such an argument is also made by M. A. Morrison in *Understanding Quantum Physics: A User's Manual*. The correct derivation of why the $k = 0$ case cannot exist in the infinite square well was originally stated by M. Bowen and J. Coster, "Infinite Square Well: A Common Mistake," *Am. J. Phys.* **49**, 80-81 (1980) with follow-up discussion in R. C. Sapp, "Ground State of the Particle in a Box," *Am. J. Phys.* **50**, 1152-1153 (1982) and L. Yinji and H. Xianhuai, "A Particle Ground State in the Infinite Square Well," *Am. J. Phys.* **54**, 738 (1986).

which tells us that $A = \sqrt{2/L}$. Therefore the wave function

$$\psi_n(x) = \sqrt{2/L}\,\sin(n\pi x/L) \quad \text{for } 0 < x < L \tag{10.9}$$

satisfies the normalization condition. We also can consider the integral

$$\int_0^L \psi_m^*(x)\psi_n(x)\,dx = \frac{Ln\cos(n\pi)\sin(m\pi)}{(m^2 - n^2)\pi} + \frac{Lm\cos(m\pi)\sin(n\pi)}{(n^2 - m^2)\pi} = 0 \tag{10.10}$$

for $m \neq n$. Therefore we can represent these two equations together as

$$\int_0^L \psi_m^*(x)\psi_n(x)\,dx = \delta_{mn}\,, \tag{10.11}$$

where δ_{mn} is the Kronecker delta which is defined such that $\delta_{nn} = 1$ and $\delta_{m\neq n} = 0$. Hence the solutions to the infinite square well are *orthogonal* and *normalized*, or *orthonormal*.

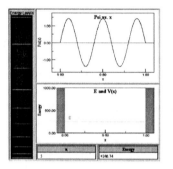

FIGURE 10.3: A quantum-mechanical particle in an infinite well. The position-space wave function, the potential energy well, and the energy levels for such a well are all shown.

In the second animation, the first 10 normalized wave functions are shown for a box with $L = 1$, along with the energy spectrum. In the animation $\hbar = 2m = 1$. You can click-drag in the energy spectrum on the left to change the energy state. As you do so, the displayed energy turns from green to red.

Note that these wave functions, $\psi_n(x)$, are only non-zero in the spatial region $0 < x < L$ and are zero everywhere else.[3] These wave functions are not so simple after all. In fact, these wave functions have a kink (a discontinuous first derivative) at $x = 0$ and $x = L$. Normally this is not acceptable for a wave function, but here the potential energy function for the infinite square well is so *badly behaved*, these wave functions are actually acceptable.

We can now also calculate expectation values of position and momentum. We find that: $\langle \hat{x} \rangle = L/2$ and $\langle \hat{p} \rangle = 0$, as expected. We also find that: $\langle \hat{x}^2 \rangle = L^2\left(\frac{1}{3} - \frac{1}{2n^2\pi^2}\right)$, $\langle \hat{p}^2 \rangle = \frac{n^2\pi^2\hbar^2}{L^2}$, and hence $\langle \hat{H} \rangle = \frac{n^2\pi^2\hbar^2}{2mL^2}$, again as expected.

---

[3]These wave functions can also be written as $\psi_n(x) = \sqrt{2/L}\,\sin(n\pi x/L)\,\Theta(x)\,\Theta(L - x)$. This representation uses two Heaviside step functions, $\Theta(\xi)$, to explicitly show the region in which the wave function is valid. This can be accomplished because the step function, $\Theta(\xi)$, is zero for $\xi < 0$ and is 1 for $\xi > 0$.

10.3 EXPLORING CHANGING WELL WIDTH

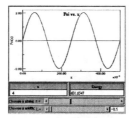

FIGURE 10.4: The energy eigenstates for a quantum-mechanical particle in an infinite well. The well width and quantum number n can be changed to see the resulting change in energy.

The animation shows position-space wave functions in an infinite square well. The width, L, of the well can be changed as can the energy eigenstate, n.

(a) What is the value of $\hbar^2/2m$ in this animation? How do you know?

(b) For what value of the width of the well can you get $E_n = n^2$?

(c) How does the energy of a given state depend on the well width? Use the *curviness* of the wave function to support your answer.

10.4 TIME EVOLUTION

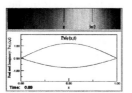

FIGURE 10.5: The time-development of an energy eigenstate in the infinite square well shown in real-and-imaginary component representation.

The time evolution of $\psi(x,t)$ is governed by the Schrödinger equation,

$$\frac{-\hbar^2}{2m}\frac{\partial^2}{\partial x^2}\psi(x,t) + V(x)\,\psi(x,t) = i\hbar\,\frac{\partial}{\partial t}\psi(x,t)\,, \qquad (10.12)$$

and we can write the solution to this equation for the energy eigenstates, $\psi_n(x)$, as

$$\psi_n(x,t) = e^{-iE_nt/\hbar}\,\psi_n(x)\,. \qquad (10.13)$$

For the infinite square well, we know that $\psi_n(x)$, given by Eq. (10.6), is an energy eigenstate which satisfies the boundary conditions and has an energy eigenvalue

E_n. Therefore the wave function becomes[4]

$$\psi_n(x,t) = e^{-iE_n t/\hbar}\,\psi_n(x) = \sqrt{\frac{2}{L}}\,e^{-i\frac{n^2\pi^2\hbar}{2mL^2}t}\,\sin\!\left(\frac{n\pi x}{L}\right), \qquad (10.14)$$

using the explicit form of the wave function and the energy.

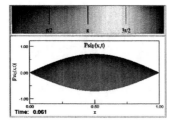

FIGURE 10.6: The time-development of an energy eigenstate in the infinite square well shown in color-as-phase representation. The color represents the phase angle.

How do we visualize this time-dependent wave function? There are several ways. We will demonstrate two. We can show the real (blue) and/or imaginary (pink) parts of the wave function which in our case are just

$$\psi_{n\,\text{Re}}(x,t) = \cos(E_n t/\hbar)\,\psi_n(x), \qquad (10.15)$$

and

$$\psi_{n\,\text{Im}}(x,t) = -\sin(E_n t/\hbar)\,\psi_n(x). \qquad (10.16)$$

Conversely, when you check the box and "input value and play," we can choose to write the wave function as

$$\psi_n(x,t) = e^{-iE_n t/\hbar}\,\psi_n(x), \qquad (10.17)$$

where we have chosen to write all of the complex behavior in an exponential (this is automatically the case here because we have written $\psi_n(x)$ as a real function). When we do this, $-E_n t/\hbar = \theta_n(t)$ is an angle in the complex plane and is called the *phase* (or phase angle) of the wave function. Note that in the case of energy eigenstates of the infinite square well, the phase of the wave function does not depend on position. We depict the amplitude of the wave function as the magnitude of the distance from the bottom to the top of the wave function at a given position and time. We represent the phase as the color of the wave function. The color strip above the animation shows the map between phase angle and color. Since quantum-mechanical time evolution involves a minus sign in the exponential, the phase evolves in time counterclockwise in the complex plane.

Explore the time dependence of the energy eigenstate of a particle in an infinite square well by changing state and representation. In the animation the time is given in terms of the time it takes the ground-state wave function to return to its original phase. In other words, $\Delta t = 1$ corresponds to an elapsed time of $2\pi\hbar/E_1$.

---

[4]This result can be derived by a Taylor series expansion of the exponential in the time-evolution operator and operate the successive powers of the Hamiltonian on the wave function, then reform the exponential.

10.5 CLASSICAL AND QUANTUM-MECHANICAL PROBABILITIES

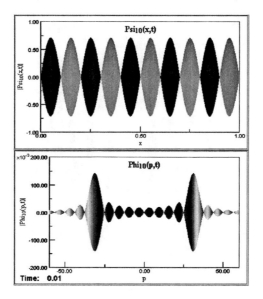

FIGURE 10.7: The time-development of an energy eigenstate in position and momentum space in the infinite square well shown in color-as-phase representation.

In the previous two Sections we discussed the energy eigenstates of the infinite square well by deriving the position-space wave function. In this Section we derive the momentum-space wave function and compare it directly to the position-space wave function as shown in the animation.

The momentum-space wave function is given by the Fourier transform of the position-space wave function. Since the position-space wave function is zero outside of the well, the Fourier transform should just involve the integral over the well: $\phi_n(p) = \frac{1}{\sqrt{2\pi\hbar}} \int_0^L \psi_n(x)\, e^{-ipx/\hbar}\, dx$, which yields:

$$\phi_n(p) = -i\sqrt{\frac{L}{4\pi\hbar}}\, e^{-ipL/2\hbar} \left[e^{+in\pi/2} \frac{\sin(\delta_{n-})}{\delta_{n-}} - e^{-in\pi/2} \frac{\sin(\delta_{n+})}{\delta_{n+}} \right], \qquad (10.18)$$

where $\delta_{n+} \equiv (pL/\hbar + n\pi)/2$ and $\delta_{n-} \equiv (pL/\hbar - n\pi)/2$. Note that the largest peaks in the momentum-space wave function occur when δ_{n+} or δ_{n-} are zero which corresponds to when $p = \pm n\pi\hbar/L$. This agrees with classical expectations. The *unexpected* structure in the momentum-space wave function arises because the position-space wave function *does not extend* over all space, thereby complicating the results of the Fourier transform.

In the animations, you can change n and see the resulting changes in the position-space and momentum-space wave functions. The time is given in terms of the time it takes the ground-state wave function to return to its original phase, *i.e.*, $\Delta t = 1$ corresponds to an elapsed time of $2\pi\hbar/E_1$.

Using the first check box, you can view the probability densities in position and momentum space. The probability density in momentum space is

$$|\phi_n(p)|^2 = \left(\frac{L}{4\hbar\pi}\right)\left[\frac{\sin^2(\delta_{n-})}{\delta_{n-}^2} + \frac{\sin^2(\delta_{n+})}{\delta_{n+}^2} - 2\cos(n\pi)\frac{\sin(\delta_{n-})\sin(\delta_{n+})}{\delta_{n-}\delta_{n+}}\right]. \quad (10.19)$$

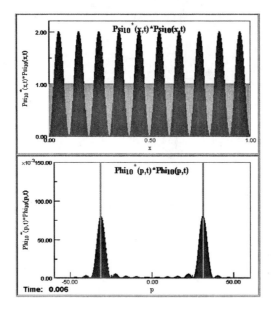

FIGURE 10.8: The quantum-mechanical probability densities in position and momentum space for the infinite square well. Superimposed on these probability densities are the corresponding normalized classical probability distributions.

In the animation you can also check the box that superimposes the normalized classical probability distributions (in pink) on the quantum-mechanical probability densities. Note that the classical position-space probability distribution is uniform over the entire well and therefore you would expect an equal likelihood of finding the classical particle anywhere in the well. The classical momentum-space probability distribution consists of two spikes at $p = \pm n\pi\hbar/L$. They correspond to the fact that half the time the classical particle is moving to the right and half the time the classical particle is moving to the left within the well.

10.6 TWO-STATE SUPERPOSITIONS

One of the simplest examples of non-trivial time-dependent states is that of an equal mix, two-state superposition in the infinite square well.[5] The position- and

---

[5]One of the earliest pedagogical visualizations of the time dependence of such a two-state system is by C. Dean, "Simple Schrödinger Wave Functions Which Simulate Classical Radiating Systems," *Am. J. Phys.* **27**, 161-163 (1959).

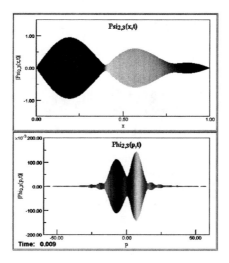

FIGURE 10.9: The time-development of an equal mix two-state (3,4) superposition in the infinite square well shown in position and momentum space.

momentum-space wave functions are just

$$\Psi_{n_1 n_2}(x,t) = \frac{1}{\sqrt{2}} \left[\psi_{n_1}(x,t) + \psi_{n_2}(x,t) \right] , \qquad (10.20)$$

and

$$\Phi_{n_1 n_2}(p,t) = \frac{1}{\sqrt{2}} \left[\phi_{n_1}(p,t) + \phi_{n_2}(p,t) \right] , \qquad (10.21)$$

where $\psi_n(x,t) = e^{-iE_n t/\hbar} \psi_n(x)$ and $\phi_n(p,t) = e^{-iE_n t/\hbar} \phi_n(p)$.

We can write these wave functions in a way that stresses their relative phases:

$$\Psi_{n_1 n_2}(x,t) = \frac{1}{\sqrt{2}} e^{-iE_{n_1} t/\hbar} \left[\psi_{n_1}(x) + e^{-i(E_{n_2} - E_{n_1})t/\hbar} \psi_{n_2}(x) \right] , \qquad (10.22)$$

and

$$\Phi_{n_1 n_2}(p,t) = \frac{1}{\sqrt{2}} e^{-iE_{n_1} t/\hbar} \left[\phi_{n_1}(p) + e^{-i(E_{n_2} - E_{n_1})t/\hbar} \phi_{n_2}(p) \right] . \qquad (10.23)$$

In this case there is a time-dependent *relative* phase that depends on the difference in energy eigenvalues and an *overall* time-dependent phase in Eqs. (10.22) and (10.23).

The animation depicts the time dependence of an arbitrary equal-mix two-state superposition. The time is given in terms of the time it takes the ground-state wave function to return to its original phase, *i.e.*, $\Delta t = 1$ corresponds to an elapsed time of $2\pi\hbar/E_1$. You can change n_1 and n_2, the default values, $n_1 = 1$ and $n_2 = 2$ represent the standard case treated in almost every textbook. Explore the time-dependent form of the position-space and momentum-space wave functions for other n_1 and n_2.

10.7   WAVE PACKET DYNAMICS

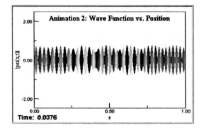

FIGURE 10.10: An initially localized Gaussian wave packet in an infinite square well. At the time shown, the wave packet has reached the so-called collapsed phase in which it is difficult to see any or the original localization.

In Section 10.6, we considered a two-state superposition within the infinite square well. We now consider a superposition of a large number of states so as to resemble an initial Gaussian wave packet at $t = 0$ in order to study the dynamics of such packets.[6]

We can examine the time dependence of such an initially localized state by choosing a Gaussian, initially localized such that is is well within the well ($0 < x < L$) and has the form

$$\Psi_G(x,0) = \frac{1}{\sqrt{\alpha\hbar\sqrt{\pi}}} \, e^{-(x-x_0)^2/2\alpha^2\hbar^2} \, e^{ip_0(x-x_0)/\hbar} \, . \tag{10.24}$$

One can show by direct calculation that $\langle x \rangle_{t=0} = x_0$, $\langle p \rangle_{t=0} = p_0$, and $\Delta x_{t=0} = \Delta x_0 = \alpha\hbar/\sqrt{2}$.

The general expression for a time-dependent wave packet solution constructed from energy eigenstates of the well is

$$\Psi_G(x,t) = \sum_{n=1}^{\infty} c_n \, e^{-iE_n t/\hbar} \, \psi_n(x) \, , \tag{10.25}$$

where the expansion coefficients satisfy $\sum_n |c_n|^2 = 1$. The expansion constants are determined by the integral

$$c_n = \int_0^L \psi_n^*(x) \, \Psi_G(x,0) \, dx \, . \tag{10.26}$$

Once we determine these coefficients, we can use Eq. (10.25) to reconstruct the wave packet and the corresponding packet dynamics. In the animation, we have two packets with $x_0 = L/2$, $\alpha = 1/10\sqrt{2}$, and one with $p_0 = 0$ and one with $p_0 = 40\pi$. We have set $\hbar = 2m = L = 1$ and the time is given in terms of the time it takes the ground-state wave function to return to its original phase, i.e., $\Delta t = 1$

---

[6]For a comprehensive review of this topic see: R. W. Robinett, "Quantum Wave Packet Revivals," *Phys. Rep.* **392**, 1-119 (2004).

corresponds to an elapsed time of $2\pi\hbar/E_1$. You can also set the starting time for the animation to study the time evolution of the packet.

The time dependence of the wave packet in an infinite square well is determined by all of the $\exp(-iE_nt/\hbar)$ factors. There are two time scales for packets within an infinite square well, they are the classical period, T_{cl}, and the revival time, T_{rev}. The classical period for the infinite square well is given by

$$T_{cl} = 2L/v_{n_0} ,$$
(10.27)

where n_0 is the mean quantum number of the packet's expansion in energy eigenstates. The analog of the classical speed, $v_{n_0} \equiv \frac{p_{n_0}}{m}$, gives a result in agreement with classical expectations. For longer time scales, we require the revival time, T_{rev}, which is given by[7]

$$T_{rev} = \frac{2\pi\hbar}{E_0} = \frac{4mL^2}{\hbar\pi} = (2n_0)T_{cl} .$$
(10.29)

Note that in the animation we have set the revival time equal to 1. For the infinite square well, no longer time scales are present due to the purely quadratic dependence of the energy eigenvalues, $E_n = n^2\pi^2\hbar^2/2mL^2$. The revival time scale can clearly be much larger than the classical period. In the second animation for example, $T_{rev}/T_{cl} = 80$.

The quantum revivals for this system are exact and any wave packet returns to its initial state after a time T_{rev}. At half this time, $t = T_{rev}/2$, the wave packet also reforms, but at a location mirrored about the center of the well and with an opposite momentum value.

At various fractional multiples of the revival time, pT_{rev}/q, where p and q are integers, the wave packet can also reform as several small *copies* (sometimes called *mini-packets* or *clones*) of the original wave packet.[8]

---

[7]In general these two time scales are calculated from the energy and they are

$$T_{cl} = \frac{2\pi\hbar}{|E'(n_0)|} \quad \text{and} \quad T_{rev} = \frac{2\pi\hbar}{|E''(n_0)|/2} ,$$
(10.28)

where the primes represent d/dn. Longer time scales can be calculated from higher-order derivatives of the energy. See for example, R. Bluhm, V. A. Kostelecký, and J. Porter, "The Evolution and Revival Structure of Localized Quantum Wave Packets," *Am. J. Phys.* **64**, 944-953 (1996).

[8]The original mathematical arguments showing how this behavior arises in general wave packet solutions were made by I. Sh. Averbukh and N. F. Perelman, "Fractional Revivals: Universality in the Long-term Evolution of Quantum Wave Packets Beyond the Correspondence Principle Dynamics," *Phys. Lett.* **A139**, 449-453 (1989) and for the infinite square well by: D. L. Aronstein and C. R. Stroud, Jr., "Fractional Wave-function Revivals in the Infinite Square Well," *Phys. Rev. A* **55**, 4526-4537 (1997).

10.8   EXPLORING WAVE PACKET REVIVALS WITH CLASSICAL ANALOGIES

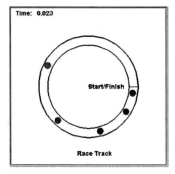

FIGURE 10.11: A classical analogy of wave packet revival behavior which depicts several racers racing around a circular track.

In Section 10.7 we saw that quantum wave packets in the infinite square well revive (reform at their original position and momentum with the exact same shape they had at $t = 0$). In Eq. (10.29) we stated the equation for the revival time, but what does this equation mean? In this Exploration we will give two ways in which we can visualize this behavior.[9]

In the first set of animations, the ones with the racers, a number of objects (cars, runners, etc.) race around a track. For simplicity, we allow the racers to pass each other by going through another racer. The angular frequency of each racer is different in a special way: the angular frequency is an integer squared times the angular speed of the slowest racer. In the second set of animations, the ones with the arrows, a number of arrows in the complex plane (phasors) indicate the phases (from each $e^{-iE_n t/\hbar}$ contribution) of the states in the infinite square well are shown. Their lengths are fixed and do not represent amplitude. The angular frequency of each phasor is different in a special way: the angular frequency is an integer squared times the angular speed of the slowest phasor. Such a depiction is often called a *phase clock*.

Answer the following questions for both the racer and phasor animations.

(a) How long does it take the slowest racer and the slowest phasor to return to its $t = 0$ position? What do these times signify?

(b) As you increase the number of racers and phasors, describe how the racers and phasors move relative to each other.

(c) Explain in your own words why a wave packet in the infinite square well revives.

---

[9]This Exploration is based in part on R. W. Robinett's talk, "Quantum Wave Packet Revivals" given at the 128th AAPT National Meeting, Miami Beach, FL, Jan. 24-28, 2004.

PROBLEMS

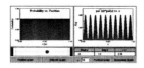

10.1. The normalized classical probability distributions and quantum-mechanical probability densities for the infinite square well are shown in position and momentum space.

    (a) Click on "Position Graph" below the right-hand graph. The graph shows the probability that a particle in the ground state is at some position x. You may vary n to see higher energy states. Under the left-hand graph, a ball is bouncing back and forth between the two walls. What does the classical probability distribution as a function of x look like? Briefly discuss your reasoning. After you answer, click "Position Graph" below the left-hand graph and check your answer. Explain why your answer agreed with, or disagreed with, the given answer.

    (b) Under what conditions *could* the right-hand graph look like the left-hand graph? In other words, what is the correspondence between the classical probability distribution and quantum position probability of a particle in a 1-d box? Check your answer using the "Position Graph" buttons.

    (c) Click on "Momentum Graph" on the right-hand graph. Displayed is a graph of the probability density in momentum space as a function of p. The box $\langle \hat{p} \rangle$ gives the expectation value of the momentum of the particle. Now click on "Velocity Graph" on the left-hand graph. What is the difference you see? Why does this difference exist?

10.2. Shown are eight position-space wave functions for infinite square wells.

    (a) Rank the wave functions in order of increasing width of the well.

    (b) Rank the wave functions in order of increasing energy.

10.3. A particle is in a one-dimensional box of length $L = 1$. The states shown are normalized. The results of the integrals that give $\langle \hat{x} \rangle$, $\langle \hat{x}^2 \rangle$, $\langle \hat{p} \rangle$, and $\langle \hat{p}^2 \rangle$ are also shown. You may vary n from 1 to 10.

    (a) What do you notice about the values of $\langle \hat{x} \rangle$ and $\langle \hat{x}^2 \rangle$ as you vary n?

    (b) What do you think $\langle \hat{x}^2 \rangle$ should become in the limit of $n \to \infty$? Why?

    (c) What do you notice about the values of $\langle \hat{p} \rangle$ and $\langle \hat{p}^2 \rangle$ as you vary n?

    (d) For $n = 1$, what are Δx and Δp?

10.4. A particle is in a superposition state in a one-dimensional box of length $L = 1$. The states shown are normalized and are an equal mix of the two states n_1 and n_2 for the infinite square well, $\Psi_{n_1 n_2}(x, t) = \frac{1}{\sqrt{2}} [\psi_{n_1}(x, t) + \psi_{n_2}(x, t)]$. Vary n_1 and n_2. The results of the integrals that give $\langle \hat{x} \rangle$, $\langle \hat{x}^2 \rangle$, $\langle \hat{p} \rangle$, and $\langle \hat{p}^2 \rangle$ are also shown. You may vary n from 1 to 10.
  (a) What do you notice about the time-independent values of $\langle \hat{x} \rangle$ and $\langle \hat{x}^2 \rangle$ as you vary n_1 and n_2?
  (b) What do you notice about the time-independent values of $\langle \hat{p} \rangle$ and $\langle \hat{p}^2 \rangle$ as you vary n_1 and n_2?

10.5. The superposition shown is an equal mix of the two states n_1 and n_2 for the infinite square well, $\Psi_{n_1 n_2}(x, t) = (1/\sqrt{2}) [\psi_{n_1}(x, t) + \psi_{n_2}(x, t)]$. The wave function evolves with time according to the Schrödinger equation. You may change states by choosing different values for n_1 and n_2. Time is shown in units of the revival time for the ground state wave function of a particle in an infinite square well. In other words, it is the time for the ground state wave function to undergo a phase change of 2π.
  (a) For $n_1 = 1$ and $n_2 = 2$, what are Δx and Δp at $t = 0$?
  (b) For $n_1 = 1$ and $n_2 = 2$, what are Δx and Δp at $t = 0.093\ (1/12)$?
  (c) For $n_1 = 1$ and $n_2 = 2$, what are Δx and Δp at $t = 0.166\ (1/6)$?
  (d) For $n_1 = 1$ and $n_2 = 2$, what are Δx and Δp at $t = 0.250\ (1/4)$?
  (e) For $n_1 = 1$ and $n_2 = 2$, what are Δx and Δp at $t = 0.333\ (1/3)$?
  (f) What do you recognize from this pattern and by looking at the wave function?
  Vary n_1 and n_2 from 1 to 10 and consider (a)-(f) for some other combination of n_1 and n_2.

10.6. The wave function shown in the upper right-hand graph, is for a particle in an infinite square well of length $L = 1$. You may import a function, $f(x)$ and check the overlap integral with the wave function. In the animation, $\hbar = 2m = 1$.
  (a) The state is not normalized. Normalize the state and report the properly normalized wave function.
  (b) How much of the total wave function is in each individual state? Report as a fraction.

C H A P T E R   11

Finite Square Wells and Other Piecewise-constant Wells

11.1 FINITE POTENTIAL ENERGY WELLS: QUALITATIVE
11.2 FINITE POTENTIAL ENERGY WELLS: QUANTITATIVE
11.3 EXPLORING THE FINITE WELL BY CHANGING WIDTH
11.4 EXPLORING TWO FINITE WELLS
11.5 FINITE AND PERIODIC LATTICES
11.6 EXPLORING FINITE LATTICES BY ADDING DEFECTS
11.7 EXPLORING PERIODIC POTENTIALS BY CHANGING WELL SEPARATION
11.8 ASYMMETRIC INFINITE AND FINITE SQUARE WELLS
11.9 EXPLORING ASYMMETRIC INFINITE SQUARE WELLS
11.10 EXPLORING WELLS WITH AN ADDED SYMMETRIC POTENTIAL
11.11 EXPLORING MANY STEPS IN INFINITE AND FINITE WELLS

INTRODUCTION

Having studied the infinite square well, in which $V = 0$ inside the well and $V = \infty$ outside the well, we now look at the bound-state solutions to other wells, both infinite and finite. The wells we will consider can be described as piecewise constant: V is a constant over a finite region of space, but can change from one region to another. We begin with the finite square well (we studied scattering-state solutions to the finite well in Section 9.6) where $V = |V_0|$ inside the well and $V = 0$ outside the well. Solutions are calculated by *piecing together* the wave function in the two regions outside the well and the one region inside the well.

11.1   FINITE POTENTIAL ENERGY WELLS: QUALITATIVE

We begin by looking at the ground state of a very deep, but not quite infinite, square well. Here the potential energy function is zero everywhere except between $-0.5 < x < 0.5$ where it is $V(x) = -|V_0|$ which is controlled by the slider. In addition, $\hbar = 2m = 1$. Notice that this energy eigenstate, here shown in position-space, resembles that of the infinite square well in shape. Drag the slider from left to right and see what happens. You should notice that as $|V_0|$ gets smaller, the wave function now *leaks* into the two classically-forbidden regions.

Now consider a shallow well by selecting the link "Show the shallow well instead." In this animation you may again drag the slider to change $|V_0|$, but you are now not limited to the ground-state wave function. To see the other bound

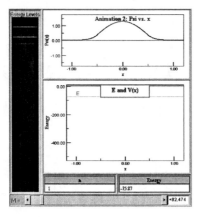

FIGURE 11.1: A shallow finite square well and its ground-state wave function.

states, simply click-drag in the energy level diagram on the left to select a level. The selected level will turn red. Again drag the slider from left to right and see what happens. You should notice that as $|V_0|$ gets smaller, the wave function now *leaks* even more into the classically-forbidden regions than before. You should also note that the number of bound states has been reduced as $|V_0|$ gets smaller. Also note that the higher n states have a larger amount of leakage than the smaller n states for the same well depth.

11.2   FINITE POTENTIAL ENERGY WELLS: QUANTITATIVE

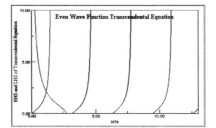

FIGURE 11.2: The left-hand side and right-hand side of Eq. (11.7) vs. ζ. The crossings of the curves that represent the LHS and RHS of Eq. (11.7) denote allowed values of ζ.

The finite square well problem is defined by a potential energy function that is zero everywhere except[1]

$$V(x) = -|V_0| \qquad -a < x < a . \tag{11.1}$$

---

[1]Note the differences between the potential energy functions describing the finite well and the infinite well. The width of the finite well is $2a$ and its walls are at $V = 0$ while the well is at $V = -|V_0|$. Bound states of the finite well, therefore, have $E < 0$. With the infinite square well, the width is L and its walls are at $V = \infty$ while the well is at $V = 0$. Bound states of the infinite well, therefore, have $E > 0$.

Since the potential energy function is finite, quantum mechanically there will be some *leakage* of the wave function into the classically-forbidden regions specified by $x > a$ and $x < -a$. We will have three regions in which we must solve the time-independent Schrödinger equation. In Region I ($x < -a$) and Region III ($x > a$), for bound-state solutions, $E < 0$, we can write the time-independent Schrödinger equation as

$$\left[\frac{d^2}{dx^2} - \kappa^2\right]\psi(x) = 0 ,$$ (11.2)

where[2] $\kappa^2 \equiv 2m|E|/\hbar^2$. Eq. (11.2) has the solutions $\psi(x) = A\exp(+\kappa x) + B\exp(-\kappa x)$. Because the wave function must be zero at $\pm\infty$ we have $\psi_{\mathrm{I}}(x) = A\exp(+\kappa x)$ and $\psi_{\mathrm{III}}(x) = D\exp(-\kappa x)$ for the solutions in Region I and III, respectively.

In Region II ($-a < x < a$), we expect an oscillatory solution since the energy is greater than the potential energy: $E > V_0$ or $|V_0| > |E|$. In this region we can write the time-independent Schrödinger as

$$\left[\frac{d^2}{dx^2} + k^2\right]\psi(x) = 0 ,$$ (11.3)

where $k^2 \equiv 2m(|V_0| - |E|)/\hbar^2$. Eq. (11.3) has the solutions $\psi_{\mathrm{II}}(x) = B\sin(kx) + C\cos(kx)$ which are valid solutions for Region II.

Next, we must match the solutions across the boundaries at $x = -a$ and $x = a$. Matching the wave functions across these boundaries means that the wave functions *and* the first derivatives of the wave functions must match at each boundary so that we have a continuous and smooth wave function (no jumps or kinks).

Since the potential energy function is symmetric about the origin, there are even *and* odd parity solutions to the bound-state problem.[3] We begin by considering the even (parity) solutions and therefore the $\psi_{\mathrm{II}}(x) = C\cos(kx)$ solution in Region II.

Matching proceeds much like the scattering cases we considered in Chapter 8. At $x = -a$ we have the conditions

$$\psi_{\mathrm{I}}(-a) = \psi_{\mathrm{II}}(-a) \quad \rightarrow \quad A\exp(-\kappa a) = C\cos(-ka) ,$$ (11.4)

and

$$\psi_{\mathrm{I}}'(-a) = \psi_{\mathrm{II}}'(-a) \quad \rightarrow \quad A\kappa\exp(-\kappa a) = -Ck\sin(-ka) .$$ (11.5)

From the symmetry in the problem, we need not consider the boundary at $x = a$ as it yields the exact same condition on energy eigenstates. We now divide the resulting two equations to give a condition for the existence of even solutions: $\kappa/k = \tan(ka)$. This is actually a constraint on the allowed energies, as both k and κ involve the energy.

---

[2]Since $E < 0$, we choose to write $E = -|E|$ to avoid any ambiguity in sign.

[3]This is due to the fact that for even potential energy functions, the Hamiltonian commutes with the parity operator. As a consequence, there are even states in which $\psi_e(-x) = \psi_e(x)$, and odd states in which $\psi_o(-x) = -\psi_o(x)$.

We now consider the following substitutions in terms of dimensionless variables:

$$\zeta \equiv ka = \sqrt{2m(|V_0| - |E|)a^2/\hbar^2}\,, \quad \text{and} \quad \zeta_0 \equiv \sqrt{2m|V_0|a^2/\hbar^2}\,, \tag{11.6}$$

where $\zeta_0 > \zeta$. Using these variables we have

$$\sqrt{(\zeta_0/\zeta)^2 - 1} = \tan(\zeta)\,. \tag{11.7}$$

This equation is a transcendental equation for ζ which itself is related via Eq. (11.6) to the energy. In addition, Eq. (11.7) only has solutions for particular values of ζ. We can solve this equation numerically or graphically, and we choose graphically in the animation. The right-hand side of Eq. (11.7) is shown in black and the left-hand side is shown in red. In the animation, $\hbar = 2m = 1$. You may also select the "Show the transcendental equation as a function of energy instead" link to see the equations as a function of energy. You can change a and $|V_0|$ by dragging the sliders to a particular value to see how the left-hand side of Eq. (11.7) changes.

We note that as the potential energy well gets shallower and/or narrower, $\zeta_0 < \pi/2$ and there exists just *one bound state*. No matter how shallow or narrow the potential energy well, there will always be at least one bound state.

As ζ_0 gets larger (meaning larger a and $|V_0|$), the number of bound-state solutions increases. In addition, the intersection of the curves on the graph approaches $\zeta = n\pi/2$, with n odd. This means that the energy (as measured from the bottom of the well) approaches that of the infinite square well of length $2a$ ($\zeta \approx n\pi/2$ yields $|V_0| - |E| = \hbar^2 k^2/2m \approx n^2\pi^2\hbar^2/2m(2a)^2$).

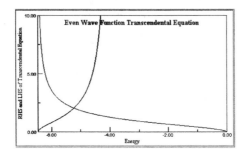

FIGURE 11.3: The left-hand side and right-hand side of Eq. (11.7) vs. energy. The crossings of the LHS and RHS denote allowed values of energy.

The solution for the odd (parity) wave functions proceeds like the even-parity case except that we use the sine solution in Region II:

$$\psi_{\mathrm{I}}(-a) = \psi_{\mathrm{II}}(-a) \quad \rightarrow \quad A\exp(-\kappa a) = B\sin(-ka)\,, \tag{11.8}$$

and

$$\psi_{\mathrm{I}}'(-a) = \psi_{\mathrm{II}}'(-a) \quad \rightarrow \quad A\kappa\exp(-\kappa a) = Bk\cos(-ka)\,. \tag{11.9}$$

Again, we need not consider the equations for $x = a$ because by symmetry, they yield the same result. We again divide the two equations to give $\kappa/k = -\cot(ka)$, and using the same substitutions in Eq. (11.6) yields: $\sqrt{(\zeta_0/\zeta)^2 - 1} = -\cot(\zeta)$.

This equation for the odd-parity solutions is shown in the animation by checking the text box and moving the slider. Note that as the potential energy well gets shallower and/or narrower, ζ_0 gets smaller, and it is possible for there to be no intersections on the graph which means that there will not be any odd-parity states. No matter how shallow or narrow the symmetric finite potential energy well, there will always be at least one bound state and it is an even-parity state.

As ζ_0 gets larger (meaning larger a and $|V_0|$), the number of bound-state solutions increases. In addition, the intersection of the curves on the graph approaches $\zeta = n\pi/2$, with n even. Again this means that the energy as measured relative to the bottom of the well approaches that of the infinite square well of length $2a$.

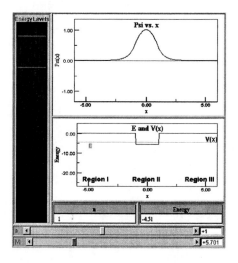

FIGURE 11.4: An adjustable (width and depth) finite square well and its ground-state wave function. Also shown are bound-state energy levels to this particular well.

In order to find the wave function, we must solve for the constants A, B, C, and D. This requires using the matching equations and then normalizing the wave function. In practice this is time consuming, instead you can view the numerical solution by clicking the "Show the wave function and well instead" link. To see other bound states, simply click-drag in the energy level diagram and select a level. The selected level will turn red.

11.3   EXPLORING THE FINITE WELL BY CHANGING WIDTH

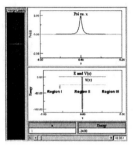

FIGURE 11.5: An adjustable-width finite square well, its ground-state wave function, and its bound-state energy levels. This particular well resembles a Dirac delta function well.

The animation shows position-space energy eigenstates for a finite square well. The half width, a, of the well can be changed with the slider as can the energy eigenstate, n, by click-dragging in the energy spectrum on the left.

(a) How does the energy spectrum depend on the well width?

(b) Describe what happens to the wave function as the well width gets smaller.

(c) Based on your observations, sketch the ground-state wave function for the bound state of an attractive Dirac delta function well.[4]

11.4   EXPLORING TWO FINITE WELLS

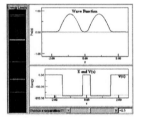

FIGURE 11.6: A set of two finite square wells and its ground-state wave function. Also shown are bound-state energy levels to this particular pair of wells.

The animation shows position-space wave functions for a pair of finite square wells. The separation, D, between the wells can be changed with the slider, as well

---

[4]If we have an attractive Dirac delta function well located at $x = 0$, $-\alpha\delta(x)$, there is an infinitely-negative spike at $x = 0$. This is an example of a *badly-behaved* potential energy function and as a consequence we expect that there will be a discontinuity in the wave function. To find how much the slope of the wave function changes (kinks) across the Dirac delta function, we integrate the Schrödinger equation near the Dirac delta function (from $-\epsilon$ to ϵ) and then let the constant $\epsilon \to 0$ at the end of the calculation:

$$\left(\frac{d\psi}{dx}\right)_{>} - \left(\frac{d\psi}{dx}\right)_{<} = -\frac{2m\alpha}{\hbar^2}\,\psi(0) \ . \tag{11.10}$$

as the energy eigenstate, n, by click-dragging in the energy spectrum on the left. In the animation, $\hbar = 2m = 1$.

(a) How does the energy spectrum depend on the individual well width?

(b) How does the energy spectrum depend on the well separation?

(c) Describe the wave functions for the default well separation.

(d) Why is it that the ground states are always even (parity)?

11.5 FINITE AND PERIODIC LATTICES

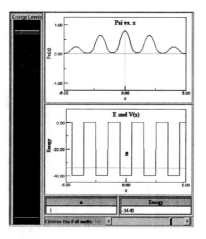

FIGURE 11.7: A set of five finite square wells (a finite lattice) and its ground-state wave function. Also shown are bound-state energy levels to this particular set of wells.

In Section 11.4 we considered what happens when we have two finite wells nearby each other. What happens when there are even more finite wells side by side? Such a situation is called a *finite lattice* of square wells. This finite lattice is modeled by a set of N finite square wells ($V_0 < 0$) each of width $b = 2a$ and a distance D apart from each other. In addition, for a finite lattice, the boundary condition on the wave function is such that it is zero at the edges of the lattice, $\psi_{\text{edges}} = 0$.

In the first animation ($\hbar = 2m = 1$), you can change the number of wells in the finite lattice from 1 to 3 to 5, while maintaining the individual well's width and depth. Notice what happens to the energy level diagram. For these particular wells, there are just two bound states possible. What happens when we increase the lattice to include three finite wells? Five finite wells? What you should notice is that the number of bound states increases as the number of wells increases. There are still two groups of states, but now each group has N individual states, where N is the number of finite wells. Therefore with three wells there are 6 bound states (three and three) and for five wells there are 10 bound states (five and five). As the number of wells increases, the number of bound states, therefore, will also increase. As the number of wells approaches the number in a metal, on the order of 10^8, the individual states form a continuous *band* of states, the individual states form a continuous band of states, while the energies between these bands are called *gaps*.

In order to consider a more quantitative model, we consider the Kronig-Penney model. In the Kronig-Penney model, the finite nature of the lattice is removed by using periodic boundary conditions: requiring the wave function at the edges of the lattice match, $\psi_{\text{left edge}} = \psi_{\text{right edge}}$. This is different than the condition we considered above. For a periodic potential, one which repeats every D, the periodicity of the potential can be expressed by $V(x) = V(x+D)$. Bloch's theorem[5] tells us that the solution to the time-independent Schrödinger equation for such a periodic potential energy function is a wave function of the form:

$$\psi(x + D) = e^{iKD}\psi(x) , \tag{11.11}$$

for a constant K. Since a solid does not go on forever, we apply the periodic boundary condition such that the wave function matches after it has gone through all N wells:

$$\psi(x + ND) = \psi(x) \tag{11.12}$$

and since $\psi(x + D) = e^{iKD}\psi(x)$, we have that

$$\psi(x + ND) = e^{iNKD}\psi(x) = \psi(x) , \tag{11.13}$$

and therefore $NKD = 2\pi n$, with $n = 0, \pm 1, \pm 2, \ldots$. In order to solve this problem

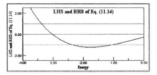

FIGURE 11.8: Shown are the left-hand and right-hand sides of Eq. (11.14) vs. energy. Only for values of the curve between the horizontal lines are there valid solutions (producing energy bands and energy gaps).

we must match wave functions and in doing so we get a transcendental equation for the bound states:

$$\cos(KD) = \cos(\sqrt{|E|\alpha})\cosh(\eta) - \frac{(2E - V_0)}{\sqrt{2(-E^2 + |E||V_0|}}\sin(\sqrt{|E|\alpha})\sinh(\eta) , \tag{11.14}$$

where $\eta = \sqrt{(|V_0| - |E|)}\beta$. The left-hand side (shown in teal in the animation) varies from 1 to -1 in tiny little steps since $KD = 2\pi n/N$, where N is a large number (the number of finite wells in the lattice). Only for certain values of the right-hand side (shown in red in the animation), between 1 to -1, are there valid solutions in the form of bands of allowable energies. This can be seen in the animation by varying the values of b, D, and $|V_0|$.

[5]For more details see pages 289-306 of R. Liboff, *Introductory Quantum Mechanics*, Addison Wesley (2003) and the original paper, F. Bloch, *Z. Physik*, **52** (1928).

11.6 EXPLORING FINITE LATTICES BY ADDING DEFECTS

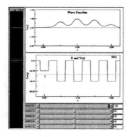

FIGURE 11.9: A finite lattice of five wells with a defect.

Shown in the animation is a finite lattice of five finite wells. Each slider controls the amount of potential energy bias you can add to each finite well. In doing so, you essentially add a defect in the finite lattice: a finite well that is not like the others. To see the other bound states simply click-drag in the energy level diagram on the left to select a level. The selected level will turn red.

(a) When all of the wells are the same, what do the energy levels of the lattice look like?

(b) Bias each one of the wells individually, both positively and negatively. What do the energy levels of the lattice look like now? How does the wave function behave in the region near the defect?

(c) Bias two of the wells in exactly the same way, either positively or negatively. What do the energy levels of the lattice look like now?

11.7 EXPLORING PERIODIC POTENTIALS BY CHANGING WELL SEPARATION

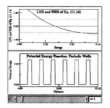

FIGURE 11.10: The left-hand and right-hand sides of Eq. (11.14) vs. energy along with a region of the periodic lattice. Only for values of the curve between the horizontal lines are there valid solutions (producing energy bands and energy gaps).

Shown in the upper graph is the transcendental equation describing the bound-state energy levels of the Kronig-Penney model. The blue lines represent the maximum of one side of the equation and the red curve represents the other side of the equation. The bottom graph shows the potential energy function: the well width $b = 2a = 1$ and the well height is $V = 6$. You may enter the well spacing, D, using the slider. In the animation, $\hbar = 2m = 1$.

(a) For the allowed changes in potential energy well spacing, describe the band structure.

(b) As the well spacing increases, what happens to the band of energies?

(c) Explain why this is the case.

11.8   ASYMMETRIC INFINITE AND FINITE SQUARE WELLS

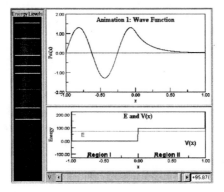

FIGURE 11.11: An asymmetric infinite square well (the infinite walls are at $x = \pm 1$) and its second excited-state wave function. Also shown are bound-state energy levels to this particular well.

This animation shows a finite potential energy well in which a constant potential energy function has been added over the right-hand side of the well. As you drag the slider to the right, the size of this *bump* or step gets larger. To see the other bound states simply click-drag in the energy level diagram on the left to select a level. The selected level will turn red. Consider Region I to be from $x = -1$ to $x = 0$ and Region II to be from $x = 0$ to $x = 1$ such that

$$V(x) = \begin{cases} +\infty & \text{for } x < -1 \\ 0 & \text{for } -1 < x < 0 \quad \text{Region I} \\ +V_0 & \text{for } 0 < x < +1 \quad \text{Region II} \\ +\infty & \text{for } +1 < x \end{cases} . \tag{11.15}$$

What happens to the wave function as we increase the step height, V_0? We begin to notice that the wave function, once having the same amplitude and curviness over both sides of the well, begins to lose this symmetry. Given the larger potential energy function in Region II, the wave function there has less curviness. In addition, the amplitude of the wave function should increase in Region II because it has a higher probability of being found there. (By simple time spent arguments: a classical particle would spend more time in Region II due to its reduced speed there.) In addition, since the added potential energy function is a constant over the entire region, the change in wave function curviness and amplitude must be uniform over Region II.

For this asymmetric infinite square well, mathematically we find that for $E <$ V_0, we have that after applying the boundary conditions at -1 and 1,

$$\psi_{\mathrm{I}}(x) = A\sin(k[x+1]) \quad \text{and} \quad \psi_{\mathrm{II}}(x) = C\sinh(\kappa[x-1]) , \qquad (11.16)$$

where $k \equiv \sqrt{2mE/\hbar^2}$ and $\kappa \equiv \sqrt{2m(V_0-E)/\hbar^2}$. Matching the two wave functions at $x = 0$ ($\psi_{\mathrm{I}}(0) = \psi_{\mathrm{II}}(0)$ and $\psi_{\mathrm{I}}'(0) = \psi_{\mathrm{II}}'(0)$) we find: $\kappa\tan(ka) = -k\tanh(\kappa b)$ which is the energy-eigenvalue equation for $E < V_0$.

Now for the $E > V_0$ case, and applying the boundary conditions at -1 and 1, we find that

$$\psi_{\mathrm{I}}(x) = A\sin(k[x+1]) \quad \text{and} \quad \psi_{\mathrm{II}}(x) = C\sin(q[x-1]) , \qquad (11.17)$$

where $k \equiv \sqrt{2mE/\hbar^2}$ and $q \equiv \sqrt{2m(E-V_0)/\hbar^2}$. Matching the two wave functions at $x = 0$, we find: $q\tan(ka) = -k\tan(qb)$ which is the energy-eigenvalue equation for $E > V_0$.

Note that for certain slider values and certain eigenstates, you may notice the same amplitude in Region I and Region II, despite the potential energy difference. This is due to the fact that the wave functions happen to match at a node.[6]

In "Animation 2" we have a finite asymmetric square well.[7] The main difference between the infinite and finite well is that there are now exponential tails in the classically forbidden regions $x < -1$ and $x > 1$.

"Animation 3" shows a well that is asymmetric in yet another way. In this case it is the sides of the well that are at different potential energies. Change the slider to see the effect of changing the height of the right side of this finite well. Does it behave in the way you might have expected?

---

[6]For more mathematical details see: M. Doncheski and R. Robinett, "Comparing Classical and Quantum Probability Distributions for an Asymmetric Infinite Well," *Eur. J. Phys.* **21**, 217-227 (2000) and "More on the Asymmetric Infinite Square Well: Energy Eigenstates with Zero Curvature," L. P. Gilbert, M. Belloni, M. A. Doncheski, and R. W. Robinett, to appear in *Eur. J. Phys.* 2005.

[7]See for example, A. Bonvalet, J. Nagle, V. Berger, A. Migus, J.-L. Martin, and M. Joffre, "Femtosecond Infrared Emission Resulting from Coherent Charge Oscillations in Quantum Wells," *Phys. Rev. Lett.* **76**, 4392-4395 (1996).

11.9 EXPLORING ASYMMETRIC INFINITE SQUARE WELLS

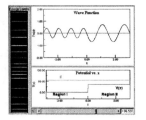

FIGURE 11.12: An asymmetric infinite square well (the infinite walls are at $x = \pm 1$) and an excited-state wave function. Also shown are bound-state energy levels to this particular well.

Shown is an infinite square well with $L = 6$ (from $x = -3$ to $x = 3$) which you can make into an asymmetric infinite square well (AISW). Consider Region I to be from $x = -3$ to $x = 0$ and Region II to be from $x = 0$ to $x = 3$. This AISW has a constant potential energy over the right half of the well (Region II) that you control with the slider. To see the other bound states simply click-drag in the energy level diagram on the left to select a level. The selected level will turn red. For the animation we have set: $\hbar = 2m = 1$. Answer the following:

(a) Make the step $V = 0$. Describe the wave function. Describe the energy spectrum.

(b) Make the step $V = 1$. What happens to the wave function in each region? What happens to the energy spectrum? What happens for large n?

(c) Make the step $V = 10$. What happens to the wave function in each region? What happens to the energy spectrum? What happens for large n?

(d) Make the step $V = 50$. What happens to the wave function in each region? What happens to the energy spectrum? What happens for small n?

11.10 EXPLORING WELLS WITH AN ADDED SYMMETRIC POTENTIAL

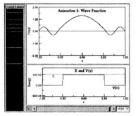

FIGURE 11.13: An infinite square well (the infinite walls are at $x = \pm 1$) with a symmetric potential energy hump, an associated excited-state wave function, and the bound-state energy levels to this well.

Shown are both an infinite and a finite square well with $L = 2$ (from $x = -1$ to $x = 1$) to which you can add a symmetric potential energy barrier or well from from $x = -0.5$ to $x = 0.5$. Use the slider to change the size of this addition. To see

the other bound states, simply click-drag in the energy level diagram on the left
to select a level. The selected level will turn red. For the animation we have set:
$\hbar = 2m = 1$. Answer the following for both the infinite and finite well:

(a) Make the step $V = -200$. What happens to the wave function for small n
and large n? What happens to the energy spectrum?

(b) Make the step $V = -100$. What happens to the wave function for small n
and large n? What happens to the energy spectrum?

(c) Make the step $V = 100$. What happens to the wave function for small n and
large n? What happens to the energy spectrum?

(d) Make the step $V = 200$. What happens to the wave function for small n and
large n? What happens to the energy spectrum?

11.11   EXPLORING MANY STEPS IN INFINITE AND FINITE WELLS

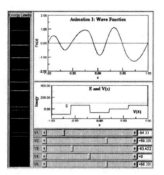

FIGURE 11.14: An infinite square well (the infinite walls are at $x = \pm 1$) with five adjustable regions of
constant potential energy, and an associated excited-state wave function. Also shown are bound-state
energy levels to this particular well.

Shown is an infinite square well with $L = 2$ (from $x = -1$ to $x = 1$) to which
you can add five potential energy steps:

V1	V2	V3	V4	V5
$-1 < x < -0.6$	$-0.6 < x < -0.2$	$-0.2 < x < 0.2$	$0.2 < x < 0.6$	$0.6 < x < 1$

by using one of the five sliders to change the size of this addition. To see the other
bound states, simply click-drag in the energy level diagram on the left to select a
level. The selected level will turn red. For the animation we have set: $\hbar = 2m = 1$.
Given the following values for the potential energies,

V1	V2	V3	V4	V5
150	0	-150	0	150
-150	0	150	0	-150

what happens to the wave function for small n and for large n? What happens to
the energy spectrum? Once you have completed this exercise, explore "Animation
2" which allows you to do something similar in a finite well.

PROBLEMS

11.1. The energy spectrum of a set of 9 finite square wells is shown. In the animation, $\hbar = 2m = 1$. To see the other bound states, simply click-drag in the energy level diagram on the left to select a level. The selected level will turn red. By using the energy level diagram, categorize each of the finite wells according to its relative depth (shallow, moderate, or deep) and relative width (narrow, moderate, or wide).

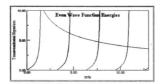

11.2. Shown in the graph is the transcendental equation describing the bound-state energy levels of finite square well. The red curve represents one side of the equation and the black curve represents the other side of the equation. In the animation, $\hbar = 2m = 1$.

(a) How many bound states are there?

(b) What are the bound state energies for the system represented by the transcendental equation?

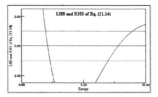

11.3. Shown in the graph is the transcendental equation describing the bound-state energy levels of a series of finite square wells. The teal lines represent the range of one side of the equation and the red curve represents the other side of the equation. In the animation, $\hbar = 2m = 1$. What are the allowed energies for the system represented by this transcendental equation?

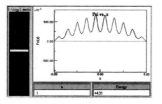

11.4. Shown is the time-independent wave function for a particle in a finite-ranged periodic potential energy function (a series of finite square potential energy wells). You may change the state by click-dragging on the energy-level diagram on the left. You may view the wave function or the probability density. In the animation, $\hbar = 2m = 1$.

(a) How many finite potential energy wells are there?

(b) Why?

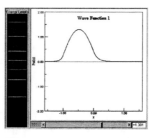

11.5. The animations show a finite potential energy well in which an unknown potential energy function is added to the well. As you drag the slider to the right, the size of this *effect* gets larger. When the slider is at 0, we recover the finite well. To see the other bound states, simply click-drag in the energy level diagram on the left to select a level. The selected level will turn red. For each animation, draw and describe what the added potential energy function does to the overall potential energy function of the finite well.

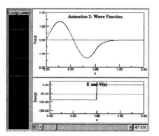

11.6. The animations show both a finite potential energy well and a half infinite-half finite potential energy well. The half well is created by splitting the original finite well in half by inserting an infinite wall at the origin. To see the other bound states, simply click-drag in the energy level diagram on the left to select a level. The selected level will turn red. Compare and contrast the wave functions and the energy levels of these two related wells.

C H A P T E R   12

Harmonic Oscillators and Other Spatially-varying Wells

12.1 THE CLASSICAL HARMONIC OSCILLATOR
12.2 THE QUANTUM-MECHANICAL HARMONIC OSCILLATOR
12.3 CLASSICAL AND QUANTUM-MECHANICAL PROBABILITIES
12.4 WAVE PACKET DYNAMICS
12.5 RAMPED INFINITE AND FINITE WELLS
12.6 EXPLORING OTHER SPATIALLY-VARYING WELLS

INTRODUCTION

In this chapter we will consider eigenstates of potential energy functions that are spatially varying, $V(x) \neq$ constant. We begin with the most recognizable of these problems, that of the simple harmonic oscillator, $V(x) = m\omega^2 x^2 / 2$. Many systems in nature *exactly* exhibit the harmonic oscillator's potential energy, but many more systems *approximately* exhibit the form of the harmonic oscillator's potential energy.[1]

12.1   THE CLASSICAL HARMONIC OSCILLATOR

We begin by considering the classical simple harmonic motion of a mass on a spring. We have chosen the mass of the ball on the spring to be 0.5 kg and the spring constant to be 2 N/m (position is given in meters and time is given in seconds). Given Hooke's law, we can write the force as $F_x = -kx$ in one dimension. This also allows us to write $F_x = ma_x = m\frac{d^2x}{dt^2} = -kx$, which using $\omega = \sqrt{k/m}$, can be written as

$$d^2x/dt^2 = -\omega^2 x \ . \tag{12.3}$$

---

[1] A generic potential energy function, $V(x)$, can be expanded in a Taylor series to yield

$$V(x) = V(x_0) + (x - x_0) \left. \frac{dV(x)}{dx} \right|_{x=x_0} + \frac{(x - x_0)^2}{2!} \left. \frac{d^2V(x)}{dx^2} \right|_{x=x_0} + \cdots . \tag{12.1}$$

If the original potential energy is symmetric about $x = 0$, we can expand about $x_0 = 0$ to yield

$$V(x) = V(0) + \frac{x^2}{2!} \left. \frac{d^2V(x)}{dx^2} \right|_{x=0} + \frac{x^4}{4!} \left. \frac{d^4V(x)}{dx^4} \right|_{x=0} + \cdots . \tag{12.2}$$

The leading non-constant term is in the form of a harmonic oscillator, and thus this potential can be approximately treated as a harmonic oscillator.

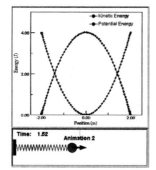

FIGURE 12.1: A classical harmonic oscillator (a mass on a spring) shown with its corresponding kinetic and potential energies plotted vs. position.

This equation has the general solution $x(t) = A\cos(\omega t) + B\sin(\omega t)$, where the coefficients A and B are determined by the initial conditions (x_0 and v_0).

"Animation 1" shows the graphs of kinetic and potential energy versus time. They have the form of $\cos^2$ (the potential energy) and the form of $\sin^2$ (the kinetic energy). We know from simple harmonic motion that if the object is initially displaced from equilibrium with no initial velocity that the solution to Eq. (12.3) is

$$x = x_0\cos(\omega t) \quad \text{and} \quad v = dx/dt = -\omega x_0\sin(\omega t). \tag{12.4}$$

Given the form of the kinetic energy (T) and the potential energy (V), we have that

$$T(t) = (kx_0^2/2)\,\sin^2(\omega t) \quad \text{and} \quad V(t) = (kx_0^2/2)\,\cos^2(\omega t). \tag{12.5}$$

"Animation 2" shows the graphs of kinetic and potential energy vs. position. The potential energy can be found from $V = -\int F \cdot dx = \frac{1}{2}kx^2 = \frac{1}{2}m\omega^2 x^2$. Since the total energy is the sum of the kinetic and the potential energies, we have that $E = p^2(t)/2m + m\omega^2 x^2(t)/2$. In the animation, the energy starts out all potential, at the equilibrium position the energy is all kinetic, and at maximum compression the energy is all potential again. The classical particle on a spring is never allowed beyond the point where all of its energy is potential (otherwise its kinetic energy would be negative), as it is classically forbidden.

12.2   THE QUANTUM-MECHANICAL HARMONIC OSCILLATOR

The solution of the one-dimensional quantum harmonic oscillator problem begins with the time-independent Schrödinger equation

$$\left[\frac{-\hbar^2}{2m}\frac{d^2}{dx^2} + \frac{1}{2}m\omega^2\hat{x}^2\right]\psi(x) = E\,\psi(x)\,, \tag{12.6}$$

where $k/2 = m\omega^2/2$. We can put this equation into a more standard form as

$$\left[\frac{d^2}{dx^2} - \frac{m^2\omega^2}{\hbar^2}\hat{x}^2 + \frac{2mE}{\hbar^2}\right]\psi(x) = 0\,. \tag{12.7}$$

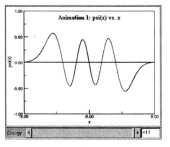

FIGURE 12.2: A quantum-mechanical harmonic oscillator's wave function.

We can further simplify the differential equation by defining the following: $\beta^2 = \frac{m\omega}{\hbar}$ and $\xi^2 = \beta^2 x^2$, which when substituted into the differential equation, gives

$$\left[\frac{d^2}{d\xi^2} - \xi^2 + \frac{2E}{\hbar\omega} \right] \psi(\xi) = 0 \; . \tag{12.8}$$

Even though this differential equation looks nothing like our original differential equation, the three terms are just the kinetic, potential, and total energies, respectively.

Consider the two special limiting cases of Eq. (12.8):

Case I: $-\xi^2 + \frac{2E}{\hbar\omega} \approx -\xi^2$, this situation results when the potential energy at a given position in the well is much greater than the total energy, $V >> E$. This is forbidden classically. We explicitly solve the differential equation

$$\left[\frac{d^2}{d\xi^2} - \xi^2 \right] \psi(\xi) = 0 \; , \tag{12.9}$$

which for large ξ yields the solutions $\psi(\xi) = e^{\pm \xi^2/2}$.

Case II: $-\xi^2 + \frac{2E}{\hbar\omega} \approx \frac{2E}{\hbar\omega}$, this corresponds to the case where the total energy is much greater than the potential energy, $E >> V$, which occurs near $x = 0$. We explicitly solve the differential equation

$$\left[\frac{d^2}{d\xi^2} + \frac{2E}{\hbar\omega} \right] \psi(\xi) = 0 \; , \tag{12.10}$$

which yields the *real* solutions $\psi(\xi) = A\cos(K\xi) + B\sin(K\xi)$ where $K^2 = \frac{2E}{\hbar\omega}$.

What does the total solution look like? In the region where $E >> V$ we have an oscillating solution. However, in the classically forbidden region $V >> E$, we have some *leakage* of the wave function into the classically-forbidden region. The well-behaved solution in this region is $e^{-\xi^2/2}$ as it goes to zero for large $\xi \to \pm\infty$.

Now that we have an idea of what the bound states should look like, we can find the entire solution to Eq. (12.8). This solution can be written as Hermite

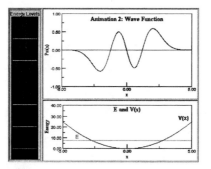

FIGURE 12.3: A harmonic oscillator and its associated ground-state wave function. Also shown are bound-state energy levels to this particular well.

polynomials weighted by a factor of $e^{-\xi^2/2} = e^{-\beta^2 x^2/2}$. These solutions can be written in terms of β and x as:

$$\psi_n(x) = A_n\,\mathcal{H}_n(\beta x)\,e^{-\beta^2 x^2/2}\,, \tag{12.11}$$

where $n = 0, 1, 2, 3, 4, \ldots$. The term $\mathcal{H}_n(\beta x)$ refers to the Hermite polynomials of order n and the A_n is the normalization factor which is equal to

$$A_n = \sqrt{\sqrt{m\omega}/(2^n\,n!\sqrt{\hbar\pi})}\,. \tag{12.12}$$

In terms of the argument ξ, the first 4 Hermite polynomials are

$$\mathcal{H}_0(\xi) = 1\,, \quad \mathcal{H}_1(\xi) = 2\xi\,, \quad \mathcal{H}_2(\xi) = 4\xi^2 - 2\,, \quad \mathcal{H}_3(\xi) = 8\xi^3 - 12\xi\,. \tag{12.13}$$

In the animations, the wave functions for a quantum harmonic oscillator are shown. The animation uses $\hbar = 2m = 1$ and $\omega = 2$. Note that in the well, the wave function's amplitude and curviness change with position. These changes are due to the fact that the potential energy function itself changes with position. The wave function is *curvier* and has a smaller amplitude nearer the center of the well as compared to the positions closer to the classical turning point.

By brute force integration we can use these wave functions to calculate the expectation values.[2] We find that $\langle \hat{x} \rangle_n = 0$ and $\langle \hat{p} \rangle_n = 0$, as expected, and

$$\langle \hat{x}^2 \rangle_n = (\hbar/2m\omega)\,(2n+1) \quad \text{and} \quad \langle \hat{p}^2 \rangle_n = (m\hbar\omega/2)\,(2n+1)\,, \tag{12.14}$$

and therefore

$$\langle E \rangle_n = \langle \hat{p}^2 \rangle_n/2m + m\omega^2 \langle \hat{x}^2 \rangle_n/2 = (n + 1/2)\hbar\omega\,. \tag{12.15}$$

This uniformity in the spacing of the harmonic oscillator energies is shown in the "spectrum" animation. To see the other bound states simply click-drag in the energy level diagram on the left to select a level. The selected level will turn red. Since we have chosen $\omega = 2$ and $\hbar = 2m = 1$, the energy spectrum is just $E_n = (2n + 1)$.

---

[2]There is an easier way. This method uses operators, called raising and lowering operators, to write the harmonic oscillator Hamiltonian. This method was pioneered by Dirac and is the easiest way to calculate expectation values.

12.3 CLASSICAL AND QUANTUM-MECHANICAL PROBABILITIES

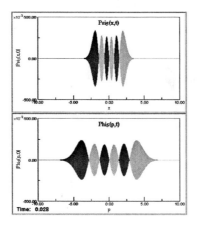

FIGURE 12.4: The position-space and momentum-space wave functions for a particle in a harmonic oscillator shown in color-as-phase representation.

In this Section we compare the energy eigenstates of the harmonic oscillator in position space to those in momentum space and then compare the resulting probability densities to their classical counterpart probability distributions.

Momentum-space wave functions can be obtained by calculating the Fourier transform of the position-space wave function (see Section 8.5). However, in the case of the harmonic oscillator, it is easier to consider the time-independent Schrödinger equation in momentum space:

$$\left[\frac{\hat{p}^2}{2m} - \frac{m\omega^2\hbar^2}{2} \frac{d^2}{dp^2} \right] \phi(p) = E\,\phi(p) . \tag{12.16}$$

In momentum space, the operator $\hat{p}$ represents the momentum and the operator $i\hbar\frac{d}{dp}$ represents the position operator, $\hat{x}$. Compare Eq. (12.6) to Eq. (12.16). What do you notice? It turns out that the two equations are in the same form, which can be seen if you make the substitution that $\hat{p} = m\omega\hat{x}$ or $\hat{x} = \hat{p}/m\omega$. Therefore, the solutions to the two differential equations are the same, apart from a scaling factor. From Eq. (12.11) and Eq. (12.16), we have that[3]

$$\phi_n(p) = B_n\,\mathcal{H}_n(\eta p)\,e^{-\eta^2 p^2/2} , \tag{12.17}$$

where $\eta = \sqrt{1/m\omega\hbar} = \beta/m\omega$. The normalization constant becomes:

$$B_n = \sqrt{1/(2^n\,n!\sqrt{m\omega\hbar\pi})} . \tag{12.18}$$

In the animations, you can change n and ω, and see the resulting changes in

---

[3]If we had Fourier transformed the position-space wave functions instead, we would have found the same result as Eq. (12.17), but multiplied by a phase $e^{in\pi/4}$ where n is the particular state's quantum number. This adds an overall phase to the momentum-space wave function and, as such, is not important.

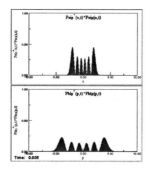

FIGURE 12.5: The position-space and momentum-space wave probability densities for a particle in a harmonic oscillator.

the position-space and momentum-space wave functions. We have used $2m = \hbar = 1$ and initially $\omega = 2$. Can you guess why we have chosen this particular value for ω? Using the first check box, you can view the probability densities in position and momentum space. In the animation, you can also check the box that superimposes the classical probability distributions (in pink) on the quantum-mechanical probability densities. Note the symmetry about $x = 0$ that the classical position-space and momentum-space probability densities exhibit.

12.4   WAVE PACKET DYNAMICS

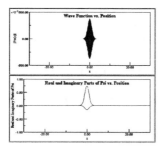

FIGURE 12.6: A quantum-mechanical wave packet in a harmonic oscillator. The wave packet is shown in both color-as-phase representation and in real-and-imaginary-component representation.

For the infinite square well, we consider a superposition of a large number of states so as to resemble an initial Gaussian wave packet at $t = 0$ in order to study the dynamics of such packets.[4] For the simple harmonic oscillator, we will outline the same approach.[5]

---

[4]For a comprehensive review of this topic see: R. W. Robinett, "Quantum Wave Packet Revivals," *Phys. Rep.* **392**, 1-119 (2004).

[5]In practice, one uses propagator methods to exactly determine the localized Gaussian wave function and its time dependence. For the details, see pages 206-208 of R. W. Robinett, *Quantum Mechanics: Classical Results, Modern Systems, and Visualized Examples*, Oxford, New York, 1997.

We can examine the time dependence of such an initially localized state by choosing a Gaussian of the form

$$\Psi_G(x,0) = \frac{1}{\sqrt{\alpha\hbar\sqrt{\pi}}} e^{-(x-x_0)^2/2\alpha^2\hbar^2} e^{ip_0(x-x_0)/\hbar} , \tag{12.19}$$

where by direct calculation: $\langle\hat{x}\rangle_{t=0} = x_0$, $\langle\hat{p}\rangle_{t=0} = p_0$, and $\Delta x_{t=0} = \Delta x_0 = \alpha\hbar/\sqrt{2}$.

The general expression for a wave packet solution constructed from such energy eigenstates is

$$\Psi_G(x,t) = \sum_{n=1}^{\infty} c_n \psi_n(x) e^{-iE_n t/\hbar} , \tag{12.20}$$

where the expansion coefficients satisfy $\sum_n |c_n|^2 = 1$. The expansion constants are determined by the integral

$$c_n = \int_{-\infty}^{\infty} \psi_n^*(x) \Psi_G(x,0) \, dx . \tag{12.21}$$

Once we determine these coefficients, we can use Eq. (12.20) to reconstruct the wave packet and study the corresponding packet dynamics. In the animation, we calculate the time dependence of packets with $x_0 = 0$ and an α, p_0, and ω that you can vary. We have set $\hbar = 2m = 1$.

The time dependence of a wave packet in a harmonic oscillator is determined by all of the $\exp(-iE_n t/\hbar)$ factors. For the harmonic oscillator there is only one characteristic time scale, $T_{cl} = 2\pi/\omega$, which gives a result in agreement with classical expectations. Because the energy of the harmonic oscillator depends linearly on the quantum number, n, there are no other time scales (unlike wave packets in the infinite square well and other wells). In fact there are only two possibilities for the wave packet's evolution: it can either be a *squeezed state* or a *coherent state*. For a squeezed state the packet remains Gaussian shaped, but as it moves throughout the well its width grows and contracts. You can see this effect by keeping the default ω and choosing $p_0 = 3$ and $\alpha = 0.5$ or $\alpha = 2$. The other situation, a coherent state (keeping the default ω and choosing $p_0 = 3$ and $\alpha = 1$) occurs for special packets and wells and is easily noticeable as the wave packet retains its exact shape throughout its motion throughout the well.

12.5   RAMPED INFINITE AND FINITE WELLS

Ramped wells consist of a potential energy function that is proportional to x added to either a finite well or an infinite well. The result is a finite or an infinite well with a ramped bottom. Solutions to such ramped wells obey a time-independent Schrödinger equation of the following form

$$\left[\frac{-\hbar^2}{2m} \frac{d^2}{dx^2} + \alpha\hat{x} \right] \psi(x) = E \, \psi(x) , \tag{12.22}$$

where α refers to the *strength* of the ramping function. We can now put this equation into a more standard form

$$\left[\frac{d^2}{dx^2} - \frac{2m\alpha}{\hbar^2}\hat{x} + \frac{2mE}{\hbar^2} \right] \psi(x) = 0 , \tag{12.23}$$

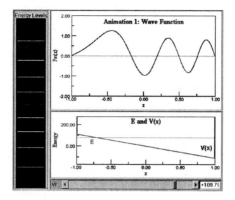

FIGURE 12.7: A ramped infinite square well and an associated excited-state wave function. Also shown are bound-state energy levels to this particular well. Note how the amplitude and curviness of the wave function change over the length of the well.

which has as its solution, Airy functions. For an infinite well, such as shown in "Animation 1," these solutions must also satisfy the boundary condition ($\psi = 0$) at the infinite wells, while in the case of a finite well, we must match the Airy functions with exponentials in the classically-forbidden regions.

Such a spatially-varying potential energy function means that for a given energy eigenstate, $E - V(x)$ will also change over the extent of the well. Two such potential energy functions (one infinite, one finite) are shown in the animation ($\hbar = 2m = 1$). Using the slider, you can change the ramping potential, V_r, to see the effect on the wave functions and the energy levels. To see the other bound states, simply click-drag in the energy level diagram on the left to select a level. The selected level will turn red.

In particular, where the well is deeper, the difference between E and V is greater. This means that the curviness of the wave function is greater there. In addition, where the well is deeper we would expect a smaller wave function amplitude.

12.6 EXPLORING OTHER SPATIALLY-VARYING WELLS

In this Exploration there are four spatially-varying wells to consider:

(a) Double harmonic oscillator well: a harmonic oscillator with $V_{added} = 10V_a e^{-10x^2}$

(b) Double anharmonic oscillator well: an anharmonic oscillator, $V(x) = V_0 x^4$, with $V_{added} = -10V_a x^2$

(c) Gaussian well: $V = -Ve^{-\alpha x^2}$

(d) Finite well with an applied electric field: $V_{added} = eE$

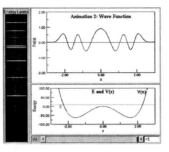

FIGURE 12.8: A double anharmonic oscillator well and an associated excited-state wave function. Also shown are bound-state energy levels to this particular well.

For each well, you may change parameters associated with the well's shape. To see the other bound states, simply click-drag in the energy level diagram on the left to select a level. The selected level will turn red. As you do so, answer the following question for each well: How does changing the variable parameter(s) change the wave functions and the energy levels?

PROBLEMS

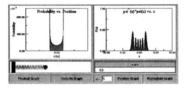
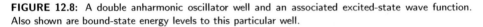

12.1. The normalized classical probability distributions and quantum-mechanical probability densities for the harmonic oscillator are shown in position and momentum space.

(a) Click on "Position Graph" below the right-hand graph. The graph shows the probability that a particle is in the ground state at some position x. You may vary n to see higher-energy states. Under the left-hand graph, a ball is attached to a spring and the spring is initially stretched. What does the classical probability distribution of finding the particle as a function of x look like? Briefly discuss your reasoning. After you answer, click "Position Graph" below the left-hand graph and check yourself. Did your answer agree with the given answer? Explain why or why not.

(b) Under what conditions would the right-hand graph look like the left-hand graph? In other words, what is the correspondence between the classical and quantum position probabilities of a particle in a harmonic oscillator potential energy function? Check your answer using the above "Position Graph" buttons.

(c) Click on "Momentum Graph" on the right-hand graph. Displayed is a graph of the probability density in momentum space as a function of p. The box $\langle p \rangle$ gives the expectation value of the momentum of the particle. Now click on "Velocity Graph" on the left-hand graph. What are the differences you see? Why do these differences exist?

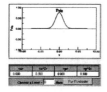

12.2. A particle is in a one-dimensional harmonic oscillator potential ($\hbar = 2m = \omega = 1$; $k = 2$). The states shown are normalized. Shown are ψ and the results of the integrals that give $\langle \hat{x} \rangle$, $\langle \hat{x}^2 \rangle$, $\langle \hat{p} \rangle$, and $\langle \hat{p}^2 \rangle$. Vary n from 1 to 10.
  (a) What do you notice about how $\langle \hat{x} \rangle$, $\langle \hat{x}^2 \rangle$, $\langle \hat{p} \rangle$, and $\langle \hat{p}^2 \rangle$ change?
  (b) Calculate $\Delta x \Delta p$ for $n = 0$. What do you notice considering $\hbar = 1$?
  (c) What is E_n? How does this agree with or disagree with the standard case for the harmonic oscillator?
  (d) How much kinetic and potential energies are in an arbitrary energy state?

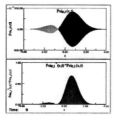

12.3. The two-state superpositions shown are an equal mix of the two states n_1 and n_2 for the harmonic oscillator. The wave function evolves with time according to the Schrödinger equation. You may change states by choosing different values for n_1 and n_2. Time is shown in units of the revival time for the ground state wave function of a particle in an infinite square well. In other words, it is the time for the ground state wave function to undergo a phase change of 2π.
  (a) For each state shown, what is the frequency of oscillation?
  (b) Are there any states shown where $\langle \hat{x} \rangle = 0$?
  (c) Are there any states shown where $\langle \hat{p} \rangle = 0$?

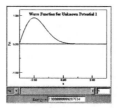

12.4. A particle is confined to a box with hard walls at $x = -3$ and $x = 3$ and an unknown potential energy function is added within the box. Change the energy slider and examine the solutions to the Schrödinger equation for this system. Examine each energy eigenfunction and sketch a potential energy function that is consistent with your observations.

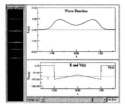

12.5. The animation shows a finite square well where a potential energy function can be added. Using the slider the bottom of the finite well can be made to vary from $\vee$ shaped to $\wedge$ shaped. To see the other bound states, simply click-drag in the energy level diagram on the left to select a level. The selected level will turn red.

Before looking at the animation:
  (a) Draw the first few energy eigenstates for the finite well with a $\vee$-shaped bottom.
  (b) Draw the first few energy eigenstates for the finite well with a $\wedge$-shaped bottom.
  (c) Once you have drawn your wave functions, use the animation to check your results. Describe in words why the wave functions behave the way they do.

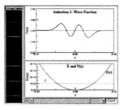

12.6. The animations show both a harmonic oscillator well and a half infinite-half harmonic oscillator well. The half well is created by splitting the original well in half by inserting an infinite wall at the origin. The slider changes the strength of the harmonic oscillator by increasing the value of k ($\hbar = 2m = 1$). To see the other bound states, simply click-drag in the energy level diagram on the left to select a level. The selected level will turn red. Compare and contrast the wave functions and the energy levels of these two related wells for $k = 2$ and $k = 4$.

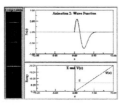

12.7. The animations show both a $\vee$-shaped potential energy well and a half infinite-half $\vee$-shaped potential energy well. The half well is created by splitting the original well in half by inserting an infinite wall at the origin. To see the other bound states, simply click-drag in the energy level diagram on the left to select a level. The selected level will turn red. Compare and contrast the wave functions and the energy levels of these two related wells.

CHAPTER   13

Multi-dimensional Wells

13.1 THE TWO-DIMENSIONAL INFINITE SQUARE WELL
13.2 TWO PARTICLES IN A ONE-DIMENSIONAL INFINITE WELL
13.3 EXPLORING SUPERPOSITIONS IN THE TWO-DIMENSIONAL INFINITE WELL
13.4 EXPLORING THE TWO-DIMENSIONAL HARMONIC OSCILLATOR
13.5 PARTICLE ON A RING
13.6 ANGULAR SOLUTIONS OF THE SCHRÖDINGER EQUATION
13.7 THE COULOMB POTENTIAL FOR THE IDEALIZED HYDROGEN ATOM
13.8 RADIAL REPRESENTATIONS OF THE COULOMB SOLUTIONS
13.9 EXPLORING SOLUTIONS TO THE COULOMB PROBLEM

INTRODUCTION

Thus far we have concerned ourselves with one-dimensional (non-relativistic) problems in quantum mechanics. We now consider the extension to systems with more than one degree of freedom in more than one dimension. Doing so extends our discussion of quantum-mechanical systems to include more real-world-like situations. We finish with the Coulomb potential which is the potential energy function responsible for basic atomic structure.

13.1   THE TWO-DIMENSIONAL INFINITE SQUARE WELL

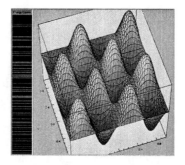

FIGURE 13.1: A two-dimensional infinite square well and an associated excited-state wave function. Also shown are bound-state energy levels to this particular well.

Recall that in one-dimension, the infinite square well confines a particle to be between 0 to L in the x direction. In this case, the time-independent Schrödinger

153

equation is

$$-\frac{\hbar^2}{2m}\frac{d^2}{dx^2}\psi(x) = E\psi(x) , \qquad (13.1)$$

which for the application of the boundary conditions gives the solutions $\psi_n(x) = \sqrt{2/L}\sin(n\pi x/L)$ and $E_n = \frac{n^2\pi^2\hbar^2}{2mL^2}$ where $n = 1, 2, 3, \ldots$.

We now seek to extend this to two dimensions for a symmetric two-dimensional infinite well, where $V(x,y) = 0$ for $-a < x < a$ and $-b < y < b$ and $V(x,y) = \infty$ elsewhere. The Hamiltonian has changed from the one-dimensional case with the addition of a term that involves the kinetic energy associated with the y direction. The wave function must therefore also be a function of both x and y and satisfy the two-dimensional time-independent Schrödinger equation:

$$\left[-\frac{\hbar^2}{2m}\frac{\partial^2}{\partial x^2} - \frac{\hbar^2}{2m}\frac{\partial^2}{\partial y^2}\right]\psi(x,y) = E\psi(x,y) . \qquad (13.2)$$

How do we solve this time-independent Schrödinger equation? We begin by using the technique of *separation of variables* to write the wave function as $\psi(x,y) = \psi(x)\psi(y)$. In doing so, we see that only the part of the wave function that is of one variable or the other gets differentiated by the kinetic energy terms. Thus we expect

$$\left[E_{n_x} + E_{n_y}\right]\psi(x)\psi(y) = E\psi(x)\psi(y) , \qquad (13.3)$$

which we can separate into

$$-\frac{\hbar^2}{2m}\frac{d^2}{dx^2}\psi(x) = E_x\psi(x) \quad \text{and} \quad -\frac{\hbar^2}{2m}\frac{d^2}{dy^2}\psi(y) = E_y\psi(y) . \qquad (13.4)$$

For the individual boundary conditions yields the solutions $\psi_{n_x}(x) = \sqrt{1/a}\sin(n_x\pi(x+1)/2a)$ and $E_{n_x} = \frac{n_x^2\pi^2\hbar^2}{8ma^2}$ and $\psi_{n_y}(y) = \sqrt{1/b}\sin(n_y\pi(y+1)/2b)$ and $E_{n_y} = \frac{n_y^2\pi^2\hbar^2}{8mb^2}$. We have for the entire solution, therefore,

$$E_{n_x n_y} = \frac{n_x^2\pi^2\hbar^2}{8ma^2} + \frac{n_y^2\pi^2\hbar^2}{8mb^2} = \frac{\pi^2\hbar^2}{8m}\left[\frac{n_x^2}{a^2} + \frac{n_y^2}{b^2}\right] , \qquad (13.5)$$

and $\psi_{n_x n_y}(x) = \sqrt{1/ab}\sin(n_x\pi(x+1)/2a)\sin(n_y\pi(y+1)/2b)$.

Consider the case where $a = b = L$ which is shown in the animation ($\hbar = 2m = 1$ and $L = 1$). You can view the wave function in either a three-dimensional plot or as a contour plot or view the probability density as a three-dimensional plot or as a contour plot. For such a *square* well, the wave function and energy simplify to: $\psi_{n_x n_y}(x,y) = (1/L)\sin(n_x\pi(x+1)/2L)\sin(n_y\pi(y+1)/2L)$ and

$$E_{n_x n_y} = \frac{\pi^2\hbar^2}{8mL^2}\left(n_x^2 + n_y^2\right) , \qquad (13.6)$$

respectively. We note that Eq. (13.6) yields the energies

$$E_{11} = 2E_1 , \quad E_{21} = E_{12} = 5E_1 , \quad E_{22} = 8E_1 , \quad E_{31} = E_{13} = 10E_1 , \quad (13.7)$$

and so forth. We therefore find degeneracies of the energy eigenvalues which is due to the geometrical symmetry of the problem. The result that some energies are degenerate is called a symmetry degeneracy.

13.2   TWO PARTICLES IN A ONE-DIMENSIONAL INFINITE WELL

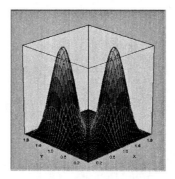

FIGURE 13.2: The probability density for two particles in an infinite square well. Shown is the antisymmetric combination.

What if we allowed *two identical particles* into a single infinite well? If they are non-interacting, the particles would move independently of each other. (They would not even collide; they would pass directly through each other!) For one particle we find that for an infinite well from $0 < x < L$, $\psi_n(x) = \sqrt{2/L}\sin(n\pi x/L)$ and $E_n = \frac{n^2\pi^2\hbar^2}{2mL^2}$. Now add another particle. The Hamiltonian has changed with the addition of terms that involve the second particle. The total wave function must now be a function of both positions x_1 and x_2:

$$\left[-\frac{\hbar^2}{2m}\frac{\partial^2}{\partial x_1^2} - \frac{\hbar^2}{2m}\frac{\partial^2}{\partial x_2^2}\right]\psi(x_1, x_2) = E\psi(x_1, x_2) . \tag{13.8}$$

Since each particle (one at x_1 and one at x_2) is independent of each other, solving this problem proceeds like the two-dimensional case in Section 13.1 and we find that $\psi_{n_1 n_2}(x_1, x_2) = (2/L)\sin(n_1\pi x_1/L)\sin(n_2\pi x_2/L)$ and

$$E_{n_1 n_2} = \frac{n_1^2\pi^2\hbar^2}{2mL^2} + \frac{n_2^2\pi^2\hbar^2}{2mL^2} = \frac{\pi^2\hbar^2}{2mL^2}\left(n_1^2 + n_2^2\right) , \tag{13.9}$$

which yields the same energy spectrum as Eq. (13.7). This degeneracy, however, is due to the *exchange* of particle one (n_1) at position x_1 with particle two (n_2) at position x_2. This is called an exchange degeneracy.

Any time we have two solutions that give the same energy, we can consider a linear combination of these solutions as another solution (since this linear solution will have the same energy as its composite states). For the case of two particles in a one-dimensional well, we combine the $\psi_{12}(x_1, x_2)$ with the $\psi_{12}(x_2, x_1)$ solution as

$$\psi_S(x_1, x_2) = \frac{1}{\sqrt{2}}\left[\psi_{12}(x_1, x_2) + \psi_{12}(x_2, x_1)\right] , \tag{13.10}$$

and

$$\psi_A(x_1, x_2) = \frac{1}{\sqrt{2}}\left[\psi_{12}(x_1, x_2) - \psi_{12}(x_2, x_1)\right] , \tag{13.11}$$

as a symmetric and an antisymmetric solution, respectively. $\psi_S(x_1, x_2)$ is symmetric under interchange because $\psi_S(x_1, x_2) = \psi_S(x_2, x_1)$. Likewise, $\psi_A(x_1, x_2)$ is antisymmetric under interchange because $\psi_A(x_1, x_2) = -\psi_A(x_2, x_1)$. We can take the square $|\psi|^2$ of each wave function and we find that these expressions are invariant to the exchange of the particles

$$|\psi_S(x_1, x_2)|^2 = |\psi_S(x_2, x_1)|^2 \quad \text{and} \quad |\psi_A(x_1, x_2)|^2 = |\psi_A(x_2, x_1)|^2 , \quad (13.12)$$

respectively, where $|\psi_S(x_1, x_2)|^2 \neq |\psi_A(x_1, x_2)|^2$.

Shown in the animation are the probability densities for two identical particles in an infinite well with $L = 2$. The x axis corresponds to the position of particle 1 and the y axis corresponds to the position of particle 2. The probability densities depicted are for one of the particles in the ground state and the other particle in the first excited state. "Animation 1" shows the symmetric solution, while "Animation 2" shows the antisymmetric solution. Note that the diagonal line across the animation ($x = y$) corresponds to $x_1 = x_2$, which is where both particles are at the same position. For the antisymmetric case, one never finds both particles at the same position (probability density is zero along this line). For the symmetric case, one is likely to find both particles at the same position (probability density is non-zero, and is actually the largest at certain points along this line).

13.3   EXPLORING SUPERPOSITIONS IN THE TWO-DIMENSIONAL INFINITE WELL

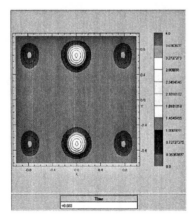

FIGURE 13.3: A contour plot of the two-state wave function (each direction) for a particle in a two-dimensional infinite square well.

One of the simplest examples of non-trivial time-dependent states is that of an equal-mix, two-state superposition in the infinite square well. Here we explore what these superpositions look like in two dimensions for a symmetric infinite square well. The individual position-space wave functions are

$$\Psi_{n_{1x} n_{2x}}(x, t) = \frac{1}{\sqrt{2}} \left[\psi_{n_{1x}}(x, t) + \psi_{n_{2x}}(x, t) \right] , \quad (13.13)$$

and

$$\Psi_{n_{1y}n_{2y}}(y,t) = \frac{1}{\sqrt{2}} \left[\psi_{n_{1y}}(y,t) + \psi_{n_{2y}}(y,t) \right] , \tag{13.14}$$

where $\Psi(x,y,t) = \Psi_{n_{1x}n_{2x}}(x,t)\Psi_{n_{1y}n_{2y}}(y,t)$, and $\psi_{n_{1x}}(x,t)$, $\psi_{n_{2x}}(x,t)$, $\psi_{n_{1y}}(y,t)$, and $\psi_{n_{2y}}(y,t)$ are the individual one-dimensional solutions.

The animation depicts the time dependence of an arbitrary equal-mix two-state superposition by showing the probability density as a three-dimensional plot and also as a contour plot. The time is given in terms of the time it takes the ground-state wave function to return to its original phase, *i.e.*, $\Delta t = 1$ corresponds to an elapsed time of $2\pi\hbar/E_1$. You can change n_{1x}, n_{2x}, n_{1y}, and n_{2y}. The default values, $n_{1x} = n_{1y} = 1$ and $n_{2x} = n_{2y} = 2$, are the two-dimensional extension of the standard one-dimensional case treated in almost every textbook, and treated here in Section 10.6.

Explore the time-dependent form of the position-space and momentum-space wave functions for other n_{1x}, n_{2x}, n_{1y}, and n_{2y}. In particular:

(a) For $n_{1x} = 1$, $n_{2x} = 2$, $n_{1y} = 1$, and $n_{2y} = 2$, describe the time dependence of the wave function. Describe the trajectory of the expectation value of position as a function of time: $\langle \hat{\mathbf{r}} \rangle = \langle \hat{x} \rangle \hat{e}_x + \langle \hat{y} \rangle \hat{e}_y$.

(b) For $n_{1x} = 1$, $n_{2x} = 2$, $n_{1y} = 1$, and $n_{2y} = 3$, describe the time dependence of the wave function. Describe the trajectory of the expectation value of position $\langle \mathbf{r} \rangle$ as a function of time.

(c) For $n_{1x} = 1$, $n_{2x} = 3$, $n_{1y} = 1$, and $n_{2y} = 3$, describe the time dependence of the wave function. Describe the trajectory of the expectation value of position $\langle \hat{\mathbf{r}} \rangle$ as a function of time.

13.4   EXPLORING THE TWO-DIMENSIONAL HARMONIC OSCILLATOR

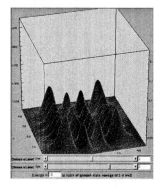

FIGURE 13.4: A two-dimensional harmonic oscillator well and an associated excited-state probability density.

In Section 12.2 we considered the one-dimensional harmonic oscillator. Here we extend that result to two dimensions: $V(x,y) = \frac{1}{2}m\omega^2(x^2 + y^2)$ (even though in general ω_x is not necessarily ω_y).

In the animations, the wave functions (three-dimensional plot and contour plot) and probability densities (three-dimensional plot and contour plot) for a two-dimensional quantum harmonic oscillator are shown. The animation uses $\hbar = 2m = 1$ and $\omega = 2$. Since we have chosen $\omega = 2$ and $\hbar = 2m = 1$, the energy spectrum for each dimension is just $E_n = (2n + 1)$ where $n = 0, 1, 2, \dots$. Hence, $E_{n_x, n_y} = 2(n_x + n_y) + 2$ where $n_x = 0, 1, 2, \dots$ and $n_y = 0, 1, 2, \dots$. Use the sliders to change the state.

(a) Change the state from $n_x = n_y = 0$ to $n_x = 0$ and $n_y = 5$. Describe the shape of the wave function.

(b) Change the state to $n_x = 5$ and $n_y = 0$. Describe the shape of the wave function. How does this wave function's shape relate to the previous wave function's shape?

(c) Change the state to $n_x = 5$ and $n_y = 5$. Describe the shape of the wave function.

(d) Describe the energy degeneracy of this system.

Do your results make sense? Try to be as complete as possible and refer back to the one-dimensional solutions.

13.5   PARTICLE ON A RING

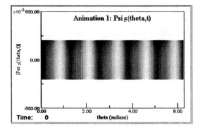

FIGURE 13.5: An energy eigenstate for a particle on a ring shown as a function of θ.

For an infinite square well potential, a particle is confined to a box of length L by two infinitely high potential energy barriers:

$$
V(x) = \begin{cases} \infty & \text{for} & x \leq 0 \\ 0 & \text{for} & 0 < x < L \\ \infty & \text{for} & x \geq L \end{cases} . \tag{13.15}
$$

For a particle on a ring (or a hoop) of radius R, the particle only has one degree of freedom, although the ring exists in two dimensions (on the xy plane). We begin by using the Schrodinger equation in cylindrical coordinates:

$$
-\frac{\hbar^2}{2m} \left[\frac{1}{\rho} \frac{\partial}{\partial \rho} \left(\rho \frac{\partial}{\partial \rho} \right) + \frac{1}{\rho^2} \frac{\partial^2}{\partial \theta^2} + \frac{\partial^2}{\partial z^2} \right] \psi(\vec{r}) + V(\vec{r})\psi(\vec{r}) = E\psi(\vec{r}) . \tag{13.16}
$$

Since the hoop is on the xy plane, $\rho = R$, $z = 0$, which are both constant. We therefore find that Eq. (13.16) reduces to

$$-\frac{\hbar^2}{2m}\left[\frac{1}{R^2}\frac{d^2}{d\theta^2}\right]\psi(\theta) = E\psi(\theta)\,, \tag{13.17}$$

since $V = 0$. Eq. (13.17) can be written in a more standard form, $\frac{d^2}{d\theta^2}\psi(\theta) + (2mR^2E/\hbar^2)\psi(\theta) = 0$, which yields the equation

$$\left[\frac{d^2}{d\theta^2} + k^2 R^2\right]\psi(\theta) = 0\,, \tag{13.18}$$

upon making the substitution, $k^2 = 2mE/\hbar^2$. The general solution is therefore $\psi(\theta) = Ae^{ikR\theta}$, where we allow k to take either positive or negative values. We next require that such solutions are valid wave functions: they must be *single valued* over the extent of the ring. This is an issue since the ring repeats every 2π. We require solutions that obey the *periodic boundary condition*, $\psi(\theta) = \psi(\theta + 2\pi)$. Therefore, it must be the case that $k = n/R$ where $n = 0, \pm 1, \pm 2, \ldots$. The *normalized* wave function is $\psi_n(\theta) = \frac{1}{\sqrt{2\pi}}e^{in\theta}$. We also find that the energy is quantized $E_n = \hbar^2 k^2/2m = n^2\hbar^2/2mR^2$, where $n = 0, \pm 1, \pm 2, \ldots$. Note that because of the periodic boundary condition, there are both positive and negative integer solutions for n, and there is a zero-energy solution ($n = 0$).

The solutions to the ring problem are shown in "Animation 1." The periodic boundary condition is already imposed on the wave functions and you can vary the quantum number n to see the effect on the wave function. In the animation, $\hbar = 2m = 1$ and the time evolution of the wave function, which is also shown in the animation, can be easily shown to obey

$$\psi_n(\theta, t) = \frac{1}{\sqrt{2\pi}}e^{-iE_n t/\hbar}e^{in\theta}\,. \tag{13.19}$$

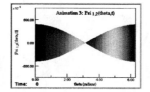

FIGURE 13.6: The time evolution of an equal-mix two-state superposition for a particle on a hoop shown as a function of θ.

Now consider "Animation 1" in which the variable is now x (as opposed to θ). In this case, x repeats every $2\pi R = L$ and the conversion between the two variables is $\theta = x/2\pi R = x/L$. Given this coordinate transformation, the wave function becomes $\psi_n(x) = \frac{1}{\sqrt{2\pi R}}e^{inx/R} = \frac{1}{\sqrt{L}}e^{2\pi inx/L}$. We also find, as before, that $E_n = n^2\hbar^2/2mR^2 = 4[n^2\pi^2\hbar^2/2mL^2]$, where $n = 0, \pm 1, \pm 2, \ldots$. Note that because of the periodic boundary condition, the energy is a factor of 4 greater than that of the usual infinite square well of the same length, there are both positive and negative integer solutions for n, and there is a zero-energy solution ($n = 0$).

In "Animation 2" you can see the time evolution of an equal-mix two-state superposition and also change the two states. Notice how some combinations (like $\{n_1 = 1, n_2 = 2\}$) yield time-dependent wave functions, while others (like $\{n_1 = 1, n_2 = -1\}$) yield time-independent standing waves.

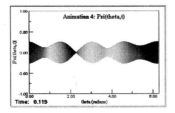

FIGURE 13.7: The time evolution of a wave packet on a hoop is shown as a function of θ.

In "Animation 3" and "Animation 4" you can see how an initial Gaussian wave packet (one without an initial momentum and one with an initial momentum) on a ring behaves. Notice the difference as compared to the wave packet in the infinite square well, Section 10.7. While the wave packet on the ring still revives like the infinite square well case, the differences in the motion are due to the lack of hard walls in the case of the ring.

Optional Visualization

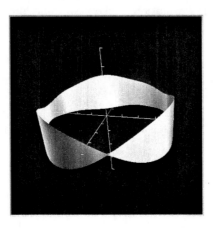

FIGURE 13.8: The Java 3D visualization of a wave packet's time evolution when it is confined to a hoop.

You may also view "Animation 3" and "Animation 4" in three dimensions if you have Java 3D installed on your computer. If you do not have Java 3D installed, go to the Sun Java 3D Web site for download: http://java.sun.com/products/java-media/3D/download.html.

13.6   ANGULAR SOLUTIONS OF THE SCHRÖDINGER EQUATION

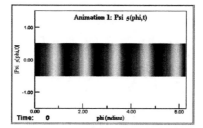

FIGURE 13.9: The ϕ part of the angular solution to the time-independent Schrödinger equation in three dimensions shown in color-as-phase representation.

Most potential energy functions in three dimensions are not often rectangular in form. In fact, they are most often in spherical coordinates (due to a spherical symmetry) and occasionally in cylindrical coordinates due to a cylindrical symmetry. We begin by considering the generalization of the time-independent Schrödinger equation to three-dimensional spherical coordinates, which is[1]

$$-\frac{\hbar^2}{2\mu}\left[\frac{1}{r^2}\frac{\partial}{\partial r}\left(r^2\frac{\partial}{\partial r}\right) + \frac{1}{r^2\sin(\theta)}\frac{\partial}{\partial\theta}\left(\sin(\theta)\frac{\partial}{\partial\theta}\right) + \frac{1}{r^2\sin^2(\theta)}\frac{\partial^2}{\partial\phi^2}\right]\psi(\vec{r})$$
$$+V(\vec{r})\psi(\vec{r}) = E\psi(\vec{r}) \ . \quad (13.20)$$

The probability per unit volume, the probability density, is $\psi^*(\vec{r})\psi(\vec{r})$ and therefore we require $\int \psi^*(\vec{r})\,\psi(\vec{r})\,d^3r = 1$ (where $d^3r = dV = r^2\sin(\theta)dr\,d\theta\,d\phi$) to maintain a probabilistic interpretation of the wave function in three dimensions.

As in the two-dimensional case, we use separation of variables variables, but now using $\psi(\vec{r}) = R(r)\mathcal{Y}(\theta,\phi)$, *i.e.*, separate the radial part from the angular part. This substitution yields

$$\left[\frac{1}{R(r)}\frac{d}{dr}\left(r^2\frac{dR(r)}{dr}\right) + \frac{1}{\mathcal{Y}\sin(\theta)}\frac{\partial}{\partial\theta}\left(\sin\theta\frac{\partial\mathcal{Y}}{\partial\theta}\right) + \frac{1}{\mathcal{Y}\sin^2(\theta)}\frac{\partial^2\mathcal{Y}}{\partial\phi^2}\right]$$
$$-\frac{2\mu r^2}{\hbar^2}[V(r) - E] = 0 \ , \quad (13.21)$$

as long as $V(\vec{r}) = V(r)$ only. Note that each term involves either r or θ and ϕ. We can separate these equations using the technique of separation of variables to give

$$\frac{1}{R(r)}\frac{d}{dr}\left(r^2\frac{dR(r)}{dr}\right) - \frac{2\mu r^2}{\hbar^2}[V(r) - E] = l(l+1) \ , \quad (13.22)$$

and

$$\frac{1}{\mathcal{Y}\sin(\theta)}\frac{\partial}{\partial\theta}\left(\sin(\theta)\frac{\partial\mathcal{Y}}{\partial\theta}\right) + \frac{1}{\mathcal{Y}\sin^2(\theta)}\frac{\partial^2\mathcal{Y}}{\partial\phi^2} = -l(l+1) \ , \quad (13.23)$$

---

[1]To avoid future confusion, we hereafter use μ for mass, and reserve m for the azimuthal (or magnetic) quantum number.

for the radial and angular parts, respectively. The constant $l(l+1)$ is the separation constant that allows us to separate one differential equation into two. We can do so because the only way for Eq. (13.21) to be true for all r, θ, and ϕ is for the angular part and the radial part to each be equal to a constant, $\pm l(l+1)$. Despite the seemingly odd form of the separation constant, it is completely general and can be made to equal any complex number.

For the angular piece, Eq. (13.23), we can again separate variables using the substitution $\mathcal{Y}(\theta, \phi) = \Theta(\theta)\Phi(\phi)$. This gives:

$$\frac{\sin(\theta)}{\Theta}\frac{d}{d\theta}\left(\sin(\theta)\frac{d\Theta}{d\theta}\right) + l(l+1)\sin^2(\theta) = m^2 , \tag{13.24}$$

and

$$\frac{1}{\Phi}\frac{d^2\Phi}{d\phi^2} = -m^2 , \tag{13.25}$$

where we have written the separation constant as $\pm m^2$, again without any loss of generality.

The $\Phi(\phi)$ part of the angular equation is a differential equation, $\frac{d^2\Phi}{d\phi^2} = -m^2\Phi$, we have solved before. We get as its unnormalized solution

$$\Phi_m(\phi) = e^{im\phi} , \tag{13.26}$$

where m is the separation constant which can be both positive and negative. Since the angle $\phi \epsilon \{0, 2\pi\}$, we must have that $\Phi_m(\phi) = \Phi_m(\phi + 2\pi)$. Like the ring problem in Section 13.5, in order for $\Phi_m(\phi)$ to be single valued means that $m = 0, \pm 1, \pm 2, \pm 3, \dots$. We show these solutions in "Animation 1."

The $\Theta(\theta)$ part of the angular equation is harder to solve. It has the unnormalized solutions

$$\Theta_l^m(\theta) = A\mathcal{P}_l^m(\cos(\theta)) , \tag{13.27}$$

where the $\mathcal{P}_l^m$ are the associated Legendre polynomials, where

$$\mathcal{P}_l^m(x) = (1 - x^2)^{|m|/2} \left(d/dx\right)^{|m|} \mathcal{P}_l(x) , \tag{13.28}$$

are calculated from the Legendre polynomials

$$\mathcal{P}_l(x) = \frac{1}{2^l l!}\left(d/dx\right)^l (x^2 - 1)^l \qquad \text{(Rodriques' formula)} . \tag{13.29}$$

The first few Legendre polynomials are

$$\mathcal{P}_0(x) = 1 , \quad \mathcal{P}_1(x) = x , \quad \text{and} \quad \mathcal{P}_2(x) = \frac{1}{2}(3x^2 - 1) , \tag{13.30}$$

or in terms of $\cos(\theta)$

$$\mathcal{P}_0 = 1 , \quad \mathcal{P}_1 = \cos(\theta) , \quad \text{and} \quad \mathcal{P}_2 = \frac{1}{2}(3\cos^2(\theta) - 1) . \tag{13.31}$$

We can also write the $\mathcal{P}_l^m(\cos(\theta))$ using the above formulas as:

$$\mathcal{P}_0^0 = 1 \; , \quad \mathcal{P}_1^1 = \sin\theta \; , \quad \mathcal{P}_1^0 = \cos\theta \; , \tag{13.32}$$

$$\mathcal{P}_2^0 = \frac{1}{2}\left(3\cos^2(\theta) - 1\right) \; , \quad \mathcal{P}_2^1 = 3\sin(\theta)\cos(\theta) \; , \quad \text{and} \quad \mathcal{P}_2^2 = 3\sin^2\theta \; . \tag{13.33}$$

We notice that $l > 0$ for Rodrigues' formula to be valid. In addition, $|m| \le l$ since $\mathcal{P}_l^{|m|>l} = 0$. (For $|m| > l$, the power of the derivative is larger than the order of the polynomial and hence the result is zero.) We also note that there must be $2l + 1$ values for m, given a particular value of l.

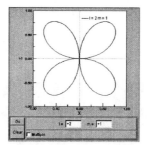

FIGURE 13.10: The θ part of the angular solution to the time-independent Schrödinger equation in three dimensions shown in the zx plane (a polar plot). The length of a vector from the origin to the wave function, $\mathcal{P}_l^m$, is the magnitude of the wave function at that angle.

Polar plots (zx plane) of associated Legendre polynomials are shown in "Animation 2." A positive angle θ is defined to be the angle down from the z axis toward the positive x axis. The length of a vector from the origin to the wave function, $\mathcal{P}_l^m$, is the magnitude of the wave function at that angle. You may vary l and m to see how $\mathcal{P}_l^m$ varies.

We normalize $\Theta_l^m(\theta)\Phi_m(\phi)$ by normalizing the angular part separately from the radial part (which we have yet to consider):

$$\int_0^{2\pi} \int_0^{\pi} \mathcal{Y}_l^{m\,*}(\theta,\phi)\,\mathcal{Y}_l^m(\theta,\phi)\,\sin(\theta)\,d\theta\,d\phi = 1 \; , \tag{13.34}$$

where $\mathcal{Y}_l^m(\theta,\phi) = \Theta_l^m(\theta)\Phi_m(\phi)$. When the $\mathcal{Y}_l^m(\theta,\phi)$ are normalized, they are called the spherical harmonics.[2]

---

[2]Classically, angular momentum is $\vec{L} = \vec{r} \times \vec{p}$. We can write $\vec{L}$ using quantum-mechanical operators in rectangular coordinates as $\hat{L}_x = \hat{y}\hat{p}_z - \hat{z}\hat{p}_y, \hat{L}_y = \hat{z}\hat{p}_x - \hat{x}\hat{p}_z$, and $\hat{L}_z = \hat{x}\hat{p}_y - \hat{y}\hat{p}_x$. We find that if we write $\hat{L}^2$ and $\hat{L}_z$ in spherical coordinates,

$$\hat{L}^2 = -\hbar^2 \left[\frac{1}{\sin(\theta)}\frac{\partial}{\partial\theta}\left(\sin(\theta)\frac{\partial}{\partial\theta}\right) + \frac{1}{\sin^2(\theta)}\frac{\partial^2}{\partial\phi^2}\right] \; , \tag{13.35}$$

and

$$\hat{L}_z = -i\hbar\frac{\partial}{\partial\phi} \; . \tag{13.36}$$

To which we note $\hat{L}^2\mathcal{Y}_l^m = l(l+1)\hbar^2\mathcal{Y}_l^m$ and $\hat{L}_z\mathcal{Y}_l^m = m\hbar\mathcal{Y}_l^m$; the spherical harmonics, the $\mathcal{Y}_l^m$, are eigenstates of $\hat{L}^2$ and $\hat{L}_z$.

The first few are

$$\mathcal{Y}_0^0(\theta, \phi) = (1/4\pi)^{1/2} \ , \tag{13.37}$$

$$\mathcal{Y}_1^{\pm 1}(\theta, \phi) = \mp (3/8\pi)^{1/2} \sin(\theta) e^{\pm i\phi} \ , \quad \mathcal{Y}_1^0(\theta, \phi) = (3/4\pi)^{1/2} \cos(\theta) \ , \tag{13.38}$$

and in general for $m > 0$,

$$\mathcal{Y}_l^m(\theta, \phi) = (-1)^m \sqrt{\tfrac{2l+1}{4\pi} \tfrac{(l-m)!}{(l+m)!}} \ e^{im\phi} \ \mathcal{P}_l^m(\cos(\theta)) \ , \tag{13.39}$$

and $\mathcal{Y}_l^{-m}(\theta, \phi) = (-1)^m \mathcal{Y}_l^{m*}(\theta, \phi)$ for $m < 0$. When we represent the spherical harmonics this way, they are automatically orthogonal:

$$\int \mathcal{Y}_l^{m*} \, \mathcal{Y}_{l'}^{m'} \, \sin(\theta) \, d\theta \, d\phi = \delta_{m,m'} \delta_{l,l'} \ . \tag{13.40}$$

Optional Visualization

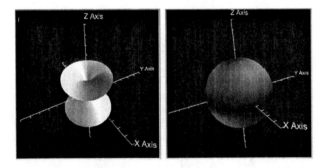

FIGURE 13.11: A Java 3D visualization of the spherical harmonics shown in two different color-as-phase representations. On the left the spherical harmonics are shown with length of a vector from the origin to the wave function as the magnitude of the wave function at that angle. On the right the magnitude of the wave function is shown by brightness.

You may also view "Animation 3" and "Animation 4" in three dimensions if you have Java 3D installed on your computer.[3] "Animation 3" shows $\mathcal{Y}_l^m(\theta, \phi)$ in the standard, but now three-dimensional, representation: The length of a vector from the origin to the wave function, $\mathcal{Y}_l^m(\theta, \phi)$, is the magnitude of the wave function at that angle. Here, the phase of the $\mathcal{Y}_l^m(\theta, \phi)$ is represented as color. "Animation 4" shows $\mathcal{Y}_l^m(\theta, \phi)$ projected on a sphere such that brightness of the wave function indicates magnitude of the wave function and color again represents the phase.

13.7   THE COULOMB POTENTIAL FOR THE IDEALIZED HYDROGEN ATOM

Now consider the radial part of the Schrödinger equation in Eq. (13.22) written as

$$\left[-\frac{\hbar^2}{2\mu} \frac{1}{r^2} \frac{d}{dr} \left(r^2 \frac{d}{dr} \right) + \frac{l(l+1)\hbar^2}{2\mu r^2} + V(r) \right] R(r) = E \, R(r) \ . \tag{13.41}$$

---

[3]If you do not have Java 3D installed, go to the Sun Java 3D Web site for download: http://java.sun.com/products/java-media/3D/download.html.

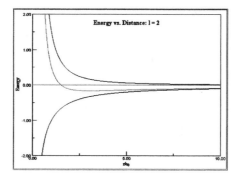

FIGURE 13.12: The effective potential for the Coulomb problem shown for $l = 2$.

We want to rewrite these terms, especially the derivative, into a more standard form. The substitution, $R(r) = u(r)/r$, simplifies the derivative term, in Eq. (13.41) to yield

$$\left[-\frac{\hbar^2}{2\mu}\frac{d^2}{dr^2} + \frac{l(l+1)\hbar^2}{2\mu r^2} + V(r) \right] u(r) = E\,u(r) \,. \tag{13.42}$$

When we find solutions to this equation, we must keep in mind that we are solving for $u(r)$, not $R(r)$, and that we must divide $u(r)$ by r to give the true radial wave function, $R(r)$. Before looking at a particular $V(r)$, we look at the general equation and interpret terms. We see what we can interpret as an *effective potential*:

$$V_{\text{eff}} = \frac{l(l+1)\hbar^2}{2\mu r^2} + V(r) \,, \tag{13.43}$$

where $\frac{l(l+1)\hbar^2}{2\mu r^2}$ is the *potential* associated with the so-called the centrifugal barrier.

Now consider the following Coulomb potential, $V = -\frac{e^2}{r}$, which describes the potential energy function for an electron in the proximity of a proton: the potential responsible for the basic structure of the hydrogen atom.[4] When we insert this Coulomb potential in the radial differential equation, we have a differential equation that describes the electron:

$$\left[-\frac{\hbar^2}{2\mu_e}\frac{d^2}{dr^2} + \frac{l(l+1)\hbar^2}{2\mu_e r^2} - \frac{e^2}{r} \right] u(r) = E\,u(r) \,, \tag{13.44}$$

where μ_e is the electron's mass and e is the charge of the electron. In "Animation 1" the effective potential for the Coulomb problem $V_{\text{eff}} = l(l+1)\hbar^2/2\mu_e r^2 - e^2/r$ is shown for $l = 0, 1, 2$. Notice that as l gets bigger, the centrifugal barrier increases as well.

To get Eq. (13.44) into standard form, divide by $-\frac{\hbar^2}{2\mu_e}$, which yields

$$\left[\frac{d^2}{dr^2} - \frac{l(l+1)}{r^2} + \frac{2\mu_e e^2}{\hbar^2 r} \right] u(r) = -\frac{2\mu_e E}{\hbar^2} u(r) \,. \tag{13.45}$$

---

[4]In Chapter 14 we discuss corrections to the Coulomb potential which are responsible for the remaining structure in the hydrogen spectral lines. We also generalize the Coulomb potential to include *hydrogenic atoms*, those with one electron and Z protons.

We now define $\kappa^2 = -\frac{2\mu_e E}{\hbar^2}$ (which is real since $E < 0$) and the dimensionless quantities $\rho = \kappa r$ and $\rho_0 = \frac{2\mu_e e^2}{\kappa \hbar^2}$. Making these substitutions yields

$$\left[\frac{d^2}{d\rho^2} - \frac{l(l+1)}{\rho^2} + \frac{\rho_0}{\rho} - 1 \right] u = 0 . \tag{13.46}$$

We begin our analysis of the solutions of this differential equation by considering the two special limiting cases:

Case I: $\rho \to 0$ $(r \to 0)$. In this case the centrifugal barrier dominates in Eq. (13.46)

$$\left[\frac{d^2}{d\rho^2} - \frac{l(l+1)}{\rho^2} \right] u = 0 . \tag{13.47}$$

We find that the general solution to this equation is $u = A\rho^{l+1} + B\rho^{-l}$ and therefore the normalizable piece is just $u \propto \rho^{l+1} \propto r^{l+1}$.

Case II: $\rho \to \infty$ $(r \to \infty)$. In this case the centrifugal term, $\frac{l(l+1)}{\rho^2}$, and the potential energy, ρ_0/ρ, vanish from Eq. (13.46) at large ρ. This leaves

$$\left[\frac{d^2}{d\rho^2} - 1 \right] u = 0 , \tag{13.48}$$

which for $E < 0$ gives the normalizable solution $u = \exp(-\rho) = \exp(-\kappa r)$.

Now that we have an idea of what the bound states should look like asymptotically, we can find the entire solution. After much algebra we first find that

$$E = -\frac{\mu_e e^4}{2n^2 \hbar^2} = -\mathcal{R}\frac{1}{n^2} , \tag{13.49}$$

where $\mathcal{R} = \frac{\mu_e e^4}{2\hbar^2}$ is the Rydberg and is 13.6 eV. This result describes the energy levels for the Coulomb problem, and hence, the basic energy level structure for the hydrogen atom. We now simplify ρ. We use $\rho = \kappa r$ and the definition of κ to find

$$\rho = \frac{\mu_e e^2}{\hbar^2 n} r = \frac{r}{na_0} , \tag{13.50}$$

where $a_0 = \hbar^2/\mu e^2$ is the Bohr radius. We can make use of further substitutions, this time yielding the radial wave functions

$$R_{nl}(r) = A_{nl}\, e^{-r/na_0} \left[\frac{(r/na_0)^{l+1}}{r} \right] v_{nl}(r/na_0) , \tag{13.51}$$

where $A_{nl} = \sqrt{(2/na_0)^3 (n-l-1)!/2n[(n+l)!]^3}$ is the normalization for the radial wave function. In addition, $v_{nl}(r/na_0) = L_{n-l-1}^{2l+1}(2r/na_0)$ are the associated Laguerre polynomials. The *unnormalized* radial wave functions, Eq. (13.51) without A_{nl}, are shown in "Animation 2." In the animation, distances are given in terms of

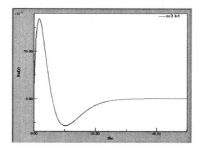

FIGURE 13.13: The radial wave function, $R_{31}(r)$, of the Coulomb problem.

Bohr radii, a_0. You may enter values of n and l and see the radial wave function that results.

We find that the entire wave function, properly normalized, is simply the product of the radial and angular solutions:

$$\psi_{nlm} = R_{nl}(r)\,\mathcal{Y}_l^m(\theta, \phi)\,. \qquad (13.52)$$

Note that given that there are n^2 states per n value, and that the energy just depends on n, the solutions have an n^2 energy degeneracy.

13.8   RADIAL REPRESENTATIONS OF THE COULOMB SOLUTIONS

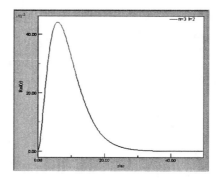
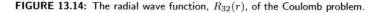

FIGURE 13.14: The radial wave function, $R_{32}(r)$, of the Coulomb problem.

In Section 13.7 we found the radial solutions

$$R_{nl}(r) = A_{nl}\, e^{-r/na_0} \left[\frac{(r/na_0)^{l+1}}{r} \right] v_{nl}(r/na_0)\,, \qquad (13.53)$$

where A_{nl} is the normalization constant and $v_{nl}(\rho) = L_{n-l-1}^{2l+1}(2\rho)$ are the associated Laguerre polynomials.

In "Animation 1" radial wave functions corresponding to the Coulomb potential, $-e^2/r$, are plotted versus distance given in Bohr radii. These are shown for $n = 1, 2, 3, 4$ with the appropriate l values. In "Animation 2," the quantum numbers are given in spectroscopic notation:

s	p	d	f
$l = 0$	$l = 1$	$l = 2$	$l = 3$

and hence $4f$ corresponds to $n = 4$ and $l = 3$. For the radial wave function, notice how the number of crossings is related to the quantum numbers n and l. You should see that the number of crossings is $n - l - 1$.

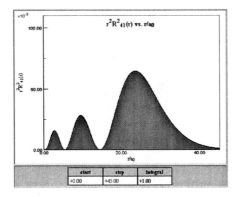

FIGURE 13.15: The radial probability density, $r^2 R_{41}^2(r)$, of the Coulomb problem is shown.

In addition, for the same quantum numbers in "Animation 2," $R_{nl}^2(r)$ and the probability density, $P_{nl}(r) = R_{nl}^2(r)r^2$, are shown. You can change the start and end of the integral for $R^2(r)$ and $R_{nl}^2(r)r^2$ as well as the range plotted in the graph by changing values and clicking the button associated with the state you are interested in. You should quickly convince yourself that while

$$\int_0^\infty R_{nl}^2(r)\, dr \neq 1 \, , \tag{13.54}$$

that

$$\int_0^\infty R_{nl}^2(r)\, r^2\, dr = \int_0^\infty u_{nl}^2(r)\, dr = 1 \, , \tag{13.55}$$

and that indeed, $P_{nl}(r) = R_{nl}^2(r)r^2$.

13.9 EXPLORING SOLUTIONS TO THE COULOMB PROBLEM

The entire solution to the Coulomb problem can be represented as

$$\psi_{nlm} = A_{nl} R_{nl}(r)\, \mathcal{Y}_l^m(\theta, \phi) \, , \tag{13.56}$$

where A_{nl} are the normalization constants, $R_{nl}(r)$ are the radial wave functions, and $\mathcal{Y}_l^m(\theta, \phi)$ are the spherical harmonics.

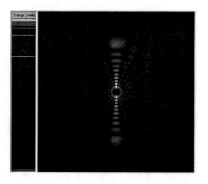

FIGURE 13.16: A polar plot (zx axis) of the entire wave function for the Coulomb problem shown in color-as-phase representation. Also shown are the energy levels for this problem.

In this Exploration, "Animation 1" depicts ψ_{nlm} in the zx plane only. In "Animation 2," $\Phi_m(\phi)$, $\mathcal{P}_l^m(\theta)$, $R_{nl}(r)$, and ψ_{nlm} are visualized in one of four panels. The entire solution, ψ_{nlm}, is visualized in the zx plane only. To generate the spherical version wave function, first imagine the rotation of ψ_{nlm} about the z axis; this gives you the shape of the wave function. Then, to get the phase of the wave function, project the phase (color) from $\Phi(\phi)$ as a function of ϕ starting with the positive x axis.

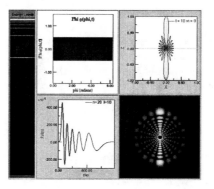

FIGURE 13.17: A plot of the entire wave function, and the ϕ, θ, and r parts of the wave function, for the Coulomb problem. Also shown are the energy levels for this particular problem.

(a) For a given value of n and l, how does the number of angular lobes in $\mathcal{P}_l^m(\theta)$ change with m?

(b) For a given value of n and l, how does the number of wavelengths (from blue to blue is one wavelength) in $\Phi_m(\phi)$ change with m?

(c) How do the non-zero l values affect the radial wave function?

(d) How do the non-zero m values affect the radial wave function?

(e) For $n = 3$, how many times does the radial wave function cross zero (change signs) for each possible value of l? Try this for a few other values for the principal quantum number, n, and see if your conclusion holds.

PROBLEMS

13.1. Shown are the solutions to an unknown potential energy function in two dimensions (rectangular). You may change the value of the slider to change the quantum number associated with each dimension. In addition, you may view the wave functions as a three-dimensional plot or as a contour plot and view the probability densities as either a three-dimensional plot or as a contour plot.

What is the potential energy well?

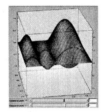

13.2. Shown are the solutions to a two-dimensional infinite square well with an added, unknown potential energy function. You may change the value of the slider to change the quantum number associated with each dimension. In addition, you may view the wave functions as a three-dimensional plot or as a contour plot and view the probability densities as either a three-dimensional plot or as a contour plot.

What is the shape and position of the added potential energy well?

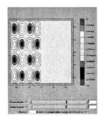

13.3. Shown are the wave functions for a single particle in a two-dimensional infinite square well. One side (x) is of length $a = 1$ and the other side (y) is of length $b = 2$. You may change the value of the slider to change the quantum number associated with each dimension. In addition, you may view the wave functions as a three-dimensional plot or as a contour plot and view the probability densities as either a three-dimensional plot or as a contour plot.

(a) Describe the energy levels as a function of n and m.

(b) What is the condition of energy degeneracy?

13.4. Wave functions for an electron in an idealized hydrogen atom (Coulomb potential) are shown up to $n = 3$. What is the relationship between n and l and the number of zero crossings of the wave function?

13.5. The probability density, $P(r) = R^2(r)r^2$, for an electron in an idealized hydrogen atom (Coulomb potential) for several states is shown as plotted versus distance.
   (a) For which state(s) do(es) the most-probable value(s) of the electron's position agree with the Bohr model?
   (b) For $1s$, $2p$, and $3d$ what is the relationship between the most-probable value and the Bohr radius?
   (c) Rank the states by the electron's most-probable position (least first).

13.6. The probability density, $P(r) = R^2(r)r^2$, for an electron in an idealized hydrogen atom (Coulomb potential) for several states is shown as plotted versus distance (Bohr radii). You can change the start and end of the integral as well as the range plotted in the graph.
   (a) For the three $n = 3$ states, find the radii at which the electron has a 50% probability of being inside and 50% outside.
   (b) The $n = 3$, $l = 0$ ($3s$) state has three regions in which the electron may be located. Find the probabilities of finding the electron in the three regions.
   (c) The $n = 3$, $l = 1$ ($3p$) state has two regions in which the electron may be located. Find the probabilities of finding the electron in the two regions.

13.7. An electron in an idealized hydrogen atom (Coulomb potential). The probability density times r, $rP(r) = R^2r^3$, integrates over all space gives the expected value for the position of the electron, represented as $\langle r \rangle$. This is shown for several states plotted versus distance given in Bohr radii. You can change the start and end of the integral as well as the range plotted in the graph.
   (a) There is an interesting relationship which relates this expected value of the radial position in terms of the Bohr theory. What is that relationship in terms of a_0?
   (b) For a given n, how does l affect the expected value of r?

PART FOUR

APPLICATIONS

C H A P T E R 14

Atomic, Molecular, and Nuclear Physics

14.1 RADIAL WAVE FUNCTIONS FOR HYDROGENIC ATOMS
14.2 EXPLORING ATOMIC SPECTRA
14.3 THE H_2^+ ION
14.4 MOLECULAR MODELS AND MOLECULAR SPECTRA
14.5 SIMPLE NUCLEAR MODELS: FINITE AND WOODS-SAXON WELLS
14.6 EXPLORING MOLECULAR AND NUCLEAR WAVE PACKETS

INTRODUCTION

Thus far we have mostly considered problems that can be solved exactly. This was not by accident. However, when discussing the real atoms, molecules, and nuclei, we must often rely on numerical techniques. In this chapter we consider a few quantum-mechanical models that have varying success in describing atoms, molecules, and nuclei.

14.1 RADIAL WAVE FUNCTIONS FOR HYDROGENIC ATOMS

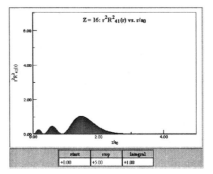

FIGURE 14.1: The radial probability density, $r^2 R_{41}^2(r)$, of the Coulomb problem with 16 protons.

In Chapter 13, we studied the Coulomb potential as applied to the idealized hydrogen atom. Here we extend this analysis to so-called hydrogenic atoms. Hydrogenic atoms are atoms with only one electron and are therefore highly ionized. We can easily extend the discussion of the radial wave functions from Section 13.8

by considering the Coulomb potential, $V = -\frac{Ze^2}{r}$, where Z represents the number of protons. We can therefore make the replacement $e^2 \rightarrow Ze^2$, in the expressions for the Bohr radius and energy equations, which give:

$$a(Z) = \frac{\hbar^2}{\mu_e Z e^2} = \frac{a_0}{Z} \, , \tag{14.1}$$

and

$$E_n(Z) = -\frac{\mu_e Z^2 e^4}{2n^2 \hbar^2} = Z^2 E_n \, , \tag{14.2}$$

where a_0 and E_n are the Bohr radius and the energy levels for the idealized hydrogen (Coulomb) problem. Notice therefore that the size of Hydrogenic atoms decreases linearly as Z increases, and the binding energy increases quadratically as Z increases.

The radial wave functions described in Eq. (13.53) change as well, such that

$$R_{Z\,nl}(r) = \sqrt{\left(\frac{2Z}{na_0}\right)^3 \frac{(n-l-1)!}{2n[(n+l)!]^3}} \, e^{-Zr/na_0} \left[\frac{(Zr/na_0)^{l+1}}{r} \right] v_{nl}(Zr/na_0) \, . \tag{14.3}$$

In the animation, these radial wave functions are shown for $Z \le 16$ and $n = 1$, 2, 3, and 4 with the appropriate (allowed) l values. The quantum numbers are given in spectroscopic notation such that $4f$ corresponds to $n = 4$ and $l = 3$. For the radial wave function, notice how the number of crossings is related to the quantum numbers n and l. You should see that the number of crossings is $n - l - 1$. Also note how the spatial extent of the radial wave functions decrease as Z increases according to Eq. (14.1).

In addition to $R_{Z\,nl}(r)$, for the same range of Z values and the same quantum numbers, $R_{Z\,nl}^2(r)$ and the probability density, $P_{Z\,nl}(r) = R_{Z\,nl}^2(r)r^2$, are also shown. You can change the start and end of the integral for $R_{Z\,nl}^2(r)$ and $R_{Z\,nl}^2(r)r^2$ as well as change the range plotted in the graph by changing values and clicking the button associated with the state in which you are interested. You should quickly convince yourself that

$$\int_0^\infty R_{Z\,nl}^2(r)\, r^2 \, dr = \int_0^\infty u_{Z\,nl}^2(r)\, dr = 1 \, , \tag{14.4}$$

and that indeed $P_{Z\,nl}(r) = R_{Z\,nl}^2(r)r^2$.

14.2 EXPLORING ATOMIC SPECTRA

In addition to the extension to hydrogenic atoms, one can also extend the Coulomb model to include effects that add *finer* structure to the atomic energy levels. These corrections include:

- The proton does not have infinite mass, $\mu_p \neq \infty$. Therefore we should replace the electron's mass, μ_e, with the reduced mass, $\mu = \frac{\mu_e \mu_p}{\mu_p + \mu_e}$.
- From the electron's point of view, the proton is moving. Moving electric charges create magnetic fields.

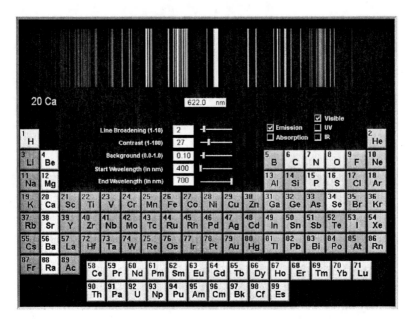

FIGURE 14.2: An interactive periodic table of the first 99 elements. The emission spectrum for Ca is also shown.

- The electron has spin angular momentum, an intrinsic magnetic moment. This, combined with the magnetic field that the electron experiences gives rise to the so-called *spin-orbit correction* to the energy.
- The electron, for large enough Z, is probably relativistic.
- The proton also has spin angular momentum.

Combining the spin-orbit interaction with the relativistic effects gives the so-called *fine-structure correction* which yields the energy levels:

$$E_{njl} = -|E_n| \left[1 + \frac{Z^2 \alpha^2}{4n^2} \left(\frac{4n}{(j + \frac{1}{2})} - 3 \right) \right], \qquad (14.5)$$

where $j = l \pm \frac{1}{2}$ and $\alpha = 1/137$ is the fine-structure constant. While these corrections are small, they do help to remove some of the degeneracies that occur in hydrogenic atom. A few corrections are shown in the following table.

The electron-proton spin interaction, due to the flipping of the electron's z component of spin, is manifest in the the 21-cm line from radio astronomy and is called the *hyperfine-structure correction*. There are also other effects from relativistic quantum mechanics and quantum field theory such as the Lamb shift.

State	Correction in $Z^2|E_n|$	J
1s	4.439945×10^{-6}	1/2
2p	-7.769904×10^{-6}	1/2
2s	-7.769904×10^{-6}	1/2
2p	-6.659917×10^{-7}	3/2
3p	-7.399908×10^{-6}	1/2
3s	-7.399908×10^{-6}	1/2
3d	-2.663967×10^{-6}	3/2
3p	-2.663967×10^{-6}	3/2
3d	-6.342778×10^{-7}	5/2
4p	-6.382421×10^{-6}	1/2
4s	-6.382421×10^{-6}	1/2
4d	-2.830465×10^{-6}	3/2
4p	-2.830465×10^{-6}	3/2
4f	-1.308198×10^{-6}	5/2
4d	-1.308198×10^{-6}	5/2
4f	-4.624943×10^{-7}	7/2

In the animation, the atomic spectra for the first 99 elements are shown. Click on an element to see its line spectra. The lines in the ultraviolet and infrared area has been given an artificial color to make them *visible*. Wavelengths are separated into three regions: ultraviolet light (100 - 400 nm), visible light (400 - 700 nm), and infrared light (700 - 47000 nm). You can measure the wavelength of the lines by putting the mouse over the spectra (some rounding errors will occur). To update changes you must click on an element once more.

14.3   THE H$_2^+$ ION

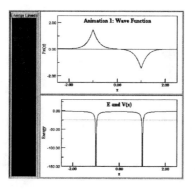

FIGURE 14.3: A one-dimensional model of an idealized H$_2^+$ ion.

One of the simplest extensions to molecules is that of the H$_2^+$ ion. The idealized H$_2^+$ ion is described by two *fixed* protons a distance r_s apart, and one electron *shared* by the two protons. The electron is a distance r_1 and r_2 away from each proton, respectively. This situation is modeled in one dimension in the animation.[1]

"Animation 1," "Animation 2," and "Animation 3" show the one-dimensional model with the protons becoming progressively closer. In each animation you can view the wave function or the probability density associated with an electron bound to the H_2^+ ion. To see the other bound states, simply click-drag in the energy level diagram on the left to select a level. The selected level will turn red. You should notice that there are two states with almost the same energy. The lowest state is *symmetric* about the origin and the next lowest-lying state is *antisymmetric* about the origin. When the protons are close together, there is a noticeable difference in the probability density associated with the symmetric and the antisymmetric state. For the symmetric state, it is relatively likely to find the electron between the two protons, while for the antisymmetric state, it is relatively unlikely to find the electron between the two protons.

We can extend this analysis to describe what happens in a neutral H_2 molecule (two electrons). In this case, shown in "Animation 4," the two wave functions in the animation represent the symmetric and antisymmetric *combinations* of the individual electron wave functions, $\Psi_S = (1/\sqrt{2})[\psi(x_1) + \psi(x_2)]$ and $\Psi_A = (1/\sqrt{2})[\psi(x_1) - \psi(x_2)]$, respectively. In the animation, the symmetric and antisymmetric spatial wave functions are shown along with $\psi(x_1)$, $\psi(x_2)$, and $-\psi(x_2)$. The total wave function (including spin) describing the two electrons must be antisymmetric upon exchange of the electrons. Therefore a symmetric spatial wave function requires an antisymmetric spin state (yielding an overall antisymmetric state) and an antisymmetric spatial wave function requires a symmetric spin state (yielding an overall antisymmetric state).

14.4   MOLECULAR MODELS AND MOLECULAR SPECTRA

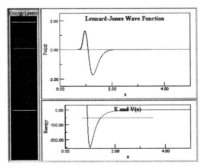

FIGURE 14.4: The Lennard-Jones potential well shown as a model of the molecular potential.

When we describe molecules, we do so by describing three energy scales: electronic, rotational, and vibrational. The electronic energy scale is the largest, on the order of the atomic energy scale of a few eV. The rotational energy scale is

---

[1]This situation is often also modeled with two finite wells that are separated. In Section 11.4, we explored this situation.

far less than than that of the electronic, but on a larger than the vibrational energy scale. Both these scales are between 100 and 1000 times smaller than the electronic energy scale.

The rotational energy scale for a diatomic molecule is modeled quite well by the three-dimensional rigid rotator. This situation can be modeled quantum mechanically by considering the three-dimensional time-independent Schrödinger equation with a fixed radius, R,

$$-\frac{\hbar^2}{2\mu}\left[\frac{1}{R^2 \sin(\theta)}\frac{\partial}{\partial\theta}\left(\sin(\theta)\frac{\partial}{\partial\theta}\right) + \frac{1}{R^2 \sin^2(\theta)}\frac{\partial^2}{\partial\phi^2}\right]\psi(\theta,\phi) = E\psi(\theta,\phi)\,, \quad (14.6)$$

which, after dividing by $-\frac{\hbar^2}{2\mu}$, looks exactly like Eq. (13.23), with the substitution $E = l(l+1)\hbar^2/2\mu R^2 = l(l+1)\hbar^2/2I_{\text{molecule}}$, where $I_{\text{molecule}} = \mu R^2$, which gives us the energy spectrum for a three-dimensional rigid rotator. The vibrational energy scale is modeled by a simple harmonic oscillator and hence $E = (n+1/2)\hbar\omega$.

In the animation, we show three important molecular potentials that closely model the rotational modes:

Kratzer Potential : $V(r) = -2V_0(\alpha/r - \alpha/2r^2)$

Morse Potential : $V(r) = V_0(e^{-2\alpha r} - 2e^{-\alpha r})$

Lennard-Jones Potential : $V(r) = V_0(\alpha/r^{12} - \alpha/r^6)$

where α is a tunable parameter. These potential energy functions are shown in "Animation 1," "Animation 2," and "Animation 3," respectively. To see the other bound states, simply click-drag in the energy level diagram on the left to select a level. The selected level will turn red.

Transitions between energy levels are also of importance and can be calculated once the energy levels are calculated. As an example of molecular spectra, shown in "Animation 4" is an approximation to the P and R branches of the CO_2 vibrational spectrum.

14.5 SIMPLE NUCLEAR MODELS: FINITE AND WOODS-SAXON WELLS

In describing nuclear material with quantum-mechanical models, there are two standard descriptions for the radial potential:[2]

The Finite Well : $V = \infty$ for $r = 0$, $V(r) = -V_0$ for $r < a$, and $V(r) = 0$ for $r > a$. The potential depth, V_0, describes the depth of the well and a describes its extent.

---

[2]Recall that in the case of a radial wave function, the time-independent Schrödinger equation for a spherically-symmetric potential is

$$\left[-\frac{\hbar^2}{2\mu}\frac{d^2}{dr^2} + \frac{l(l+1)\hbar^2}{2\mu r^2} + V(r)\right]u(r) = E\,u(r)\,, \quad (14.7)$$

where μ is the mass and $u(r) = rR(r)$. Note also that: $V_{\text{eff}}(r) = V(r) + l(l+1)/2\mu r^2$.

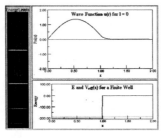

FIGURE 14.5: A finite square well with one infinite wall at $r = 0$ shown as a model of the nuclear potential.

The Woods-Saxon Potential : $V = \infty$ for $r = 0$ and $V(r) = V_0/(1 + e^{(x-a)/t})$ for $r > 0$. The potential depth, V_0, describes the depth of the well, a describes its extent, and t describes the *thickness* of the potential. This thickness describes a distance associated with the potential energy going from V_0 to 0. The shape of the Woods-Saxon potential is due to the gradual change in the charge density of the nucleus from ρ to 0, which occurs over an associated distance, t, the surface thickness.

These models, and their solutions ($u(r)$, $R(r)$, and $r^2 R^2(r) = u^2(r)$) are shown in "Animation 1" for the finite well and "Animation 2" for the Woods-Saxon potential. The angular momentum quantum number, l, can be changed and the effect can be seen on the effective potential and the resulting wave functions. To see the other bound states, simply click-drag in the energy level diagram on the left to select a level. The selected level will turn red.

14.6 EXPLORING MOLECULAR AND NUCLEAR WAVE PACKETS

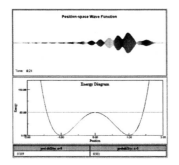

FIGURE 14.6: A wave packet in a one-dimensional double well. Such a double well reasonably represents the potential experienced by a nitrogen atom in an ammonia molecule.

This Exploration shows the *same* initial Gaussian wave packet in either a double anharmonic oscillator well or a finite well created to depict molecular or nuclear wave packets, respectively. Specifically,

One-dimensional Double Well : An anharmonic oscillator with an added negative harmonic oscillator piece, $V(x) = V_0x^4 - 10V_1x^2$. This double well can be used as a model for the ammonia molecule, NH_3. In this molecule, the three hydrogen atoms form an equilateral triangle and the nitrogen atom oscillates through the plane of the hydrogen atoms forming a pyramid shape. The potential energy function the nitrogen experiences in its oscillations is modeled relatively well by this double well potential.

Radial Finite Well : A finite well with a Coulomb *tail*, $V(r) = -V_0$ for $r < a$ and $V(r) \propto 1/r$ for $r > a$. This well depicts a model for an alpha particle in a nucleus.

(a) For the molecular wave packet animation, what happens to the packet over time? What do you notice about how the wave function behaves when $E < V$ and $E > V$? What do you notice about the probability for $x < 0$ and $x > 0$?

(b) For the nuclear wave packet animation, what happens to the packet over time? What do you notice about the probability for $x < 2$ and $x > 2$? Can you extrapolate to what will happen to the probability of $x > 2$ over time?

PROBLEMS

14.1. Angular momentum values for an electron are shown projected on the $z - x$ plane (only $+x$ projection is shown and the y axis is into the computer screen). The units of the x and z axes are in terms of $h/2\pi$ ($\hbar$). When the z component of angular momentum is greatest, what is the x component of the angular momentum?

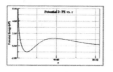

14.2. A silver atom (with total angular momentum equal to zero and $s = 1/2$ due to the valence electron) leaves an oven and passes through a non-uniform magnetic field and hits a screen as shown. What is the m_s value of the atom?

14.3. Two potential energy curves are shown (given in eV and distance is given in Bohr radii) for two different molecules. Determine which molecule requires the greater amount of energy to be disassociated and find that energy.

14.4. The potential energy curves are shown (PE given in eV and distance is given in Bohr radii) for two diatomic molecules with the same reduced mass.
(a) What part of these curves relates to the moment of inertia and k_{eff}?
(b) Determine which molecule has the larger moment of inertia and larger k_{eff}.

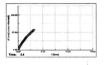

14.5. The animation depicts two 100-gram spheres connected by a spring with $k = 200 N/m$ (position given in meters and time is given in seconds). Determine the quantum number associated with the state shown.

14.6. The animation depicts two 100-gram spheres connected by a spring with $k = 200$ N/m (position given in meters and time is given in seconds). Determine the quantum number associated with the angular momentum state shown.

14.7. A source of decaying parent nuclei is placed near a detector and the total number of counts detected (black data series) are shown (time is given in hours). Determine the time constant. What fraction of the parent nuclei remain when t $= \tau$?

14.8. Shown are 100 radioactive nuclei. All nuclei have the same probability to decay within the next moment of time. A graph showing the number of not-yet-decayed nucleus N as function of time τ, is called a decay curve. If you let the nucleus above decay a few times you'll notice that the decay curve very much follows this function. The graph fits the most for large amount of radioactive nucleus present. For all radioactive sources the law of decay applies: The number of not-yet-decayed nuclei falls exponentially with time. Determine the half-life period for the decay.

C H A P T E R   15

Statistical Mechanics

15.1 EXPLORING FUNCTIONS: $g(\varepsilon)$, $f(\varepsilon)$, AND $n(\varepsilon)$
15.2 ENTROPY AND PROBABILITY
15.3 UNDERSTANDING PROBABILITY DISTRIBUTIONS
15.4 EXPLORING CLASSICAL, BOSE-EINSTEIN, AND FERMI-DIRAC STATISTICS
15.5 STATISTICS OF AN IDEAL GAS, A BLACKBODY, AND A FREE ELECTRON GAS
15.6 EXPLORING THE EQUIPARTITION OF ENERGY
15.7 SPECIFIC HEAT OF SOLIDS

INTRODUCTION

Statistical mechanics is the study of systems with large numbers of particles which serves to connect microscopic properties of individual particles to macroscopically observable quantities. This chapter limits its focus of statistical mechanics to a comparison between the classical and quantum statistics and some of the associated applications. Quantum statistics operates when particles are indistinguishable and the type of statistics depends on the spin of the particles: Bose-Einstein statistics for bosons (integer or zero spin particles) and Fermi-Dirac statistics for fermions (half-integer spin particles).

15.1   EXPLORING FUNCTIONS: $g(\varepsilon)$, $f(\varepsilon)$, AND $n(\varepsilon)$

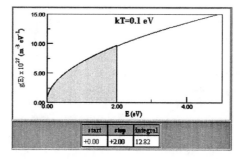

FIGURE 15.1: The number of states as a function of energy at $k_B T = 0.1$ eV.

A statistical system (whether classical or quantum mechanical) can be described in terms of the number of particles with a given energy, $n(\varepsilon)$. In general,

183

this function is the product of two functions: $n(\epsilon) = g(\varepsilon)f(\epsilon)$. $g(\varepsilon)$ counts the number of states of a given energy and is also called the density of states. $f(\varepsilon)$ is the probable distribution of the particles through the states and is also called the occupancy. The form of these functions varies depending on the system of interest: whether governed by classical statistics or quantum statistics.

The animation shows you these distribution functions for a system of fermions (since these are fermions, the correct statistics is Fermi-Dirac) at a particular temperature. You can evaluate the integral of each function from any starting point and end point. For example, in order to find the total number of particles, you must integrate $n(\varepsilon)$ over the entire energy range. How many particles are in this system? Over what energy range are states filled (a probability of 1)? Over what range are they empty (a probability of 0)? For this system, is the degeneracy constant as a function of energy?

15.2   ENTROPY AND PROBABILITY

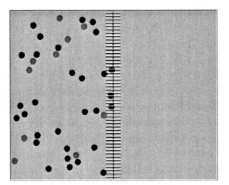

FIGURE 15.2: Two containers are separated by a membrane with particles initially in one container only.

In the animation two containers are separated by a membrane. Initially, no particles can cross the membrane. Note that the red and the blue particles are identical (they are colored so you can keep track of them). Once the particles are evenly distributed in the left chamber, you are ready to let particles through. Try "letting particles through the membrane." This animation allows about every other particle that hits the membrane to get through (equally in either direction). When there are about the same number of particles on both the left and right sides, pause the animation and count the number of red particles on each side and the number of blue particles on each side. Let the animation continue and stop it again a few seconds later when there are about the same number of particles on each side. Again, find the number of red and blue particles on each side.

(a) Given that there are 30 blue particles and 10 red particles total, if you made many such measurements, what would you expect the average number of red and blue particles to be on each side (when there are a total of 20 particles on each side)?

(b) Now, restart the animation. Once the particles are evenly distributed in the left chamber, try letting particles through the membrane a different way. Again, this animation allows about every other particle through the membrane.

(c) When there are about the same number of particles on each side, count the red and blue particles on each side. What is different about the way this membrane is set up?

(d) Is it possible for the first membrane to have this outcome? Is this outcome likely?

The reason that the second membrane does not appear *natural* is the second law of thermodynamics. One version of the second law states that the entropy of an isolated system always stays the same or increases. Entropy is a measure of the number of possible arrangement of particles or a measure of the number of microstates available to a system. This is, in some sense, a measure of *disorder* in the system. Hence, natural systems tend towards greater disorder. In the animations the first membrane seems natural because it allows for the most disorder—a random distribution of reds and blues on both sides. This is compared to the second membrane that only allows blue particles through, and thus the right side will always have only blue particles in it.

Another way to interpret the second law of thermodynamics is in terms of probability. It is possible with the first animation to get 0 red particles in the right chamber, but it is very unlikely (just like it is possible you will win the lottery, but it is very unlikely). Thus, it is possible that there is no difference between the membrane in the first and second animations, but it is highly unlikely. Consider the animation above with only six particles: four blue and two red. To keep track of the particles, some are different shades of blue and some are different shades of red (this is a model of classical, not quantum mechanical, particles because they are distinguishable). Run the animation and notice how often there are three blue particles on the right side when there are three particles on each side. What follows will allow you to calculate the probability of this happening and show that when there are three particles on each side there is a 20% chance that there will be three blue ones in the right chamber.

(e) Considering the different arrangements of three particles on each side, note that there are four different ways to get three blues on the right and two reds and one blue on the left (list these and click here to show them). Similarly, there are the same four ways to get three blue particles in the left chamber.

(f) There are six ways to get the light red on the left and the dark red on the right with two blues each (click here to show these). Again, there are the same six ways to have the dark red on the left and the light red on the right.

(g) This gives a total of how many different arrangements (of three particles on each side)? Since all these states are equally likely, you have only a 20% chance of having three blues in the right chamber.

As we add more particles, it becomes less likely to get all of one color on one side. With 40 particles, 30 blue and 10 red, there is only around a 0.02%

chance that when there are 20 particles on each side, there will be 20 blue ones on the left and 10 red and 10 blue on the right. This is not impossible, but not very likely (better odds than your local lottery, where your odds might be around one in a couple of million). A more ordered state (20 blues on the right) is less likely, statistically, than a less ordered state (reds on both sides of the membrane, which is a more even mixing). Entropy is related to the number of available states that correspond to a given arrangement (mathematically, $S = k_B \ln W$, where S is entropy, W is the number of equivalent arrangements or microstates, and k_B is the Boltzmann constant). Section 15.3 allows you to change particle arrangements and calculates the corresponding microstates, and thus, the most probable particle distribution for both classical systems (like this) and quantum-mechanical systems of bosons and fermions.

15.3 UNDERSTANDING PROBABILITY DISTRIBUTIONS

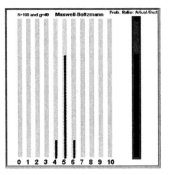

FIGURE 15.3: The distribution of particles for a system of 11 boxes, labeled 0 to 10.

The best distribution of many particles across regions of phase space is that which maximizes the number of ways the particles can be arranged in these regions (or *boxes* for the purpose of this animation) for a given energy and given number of particles. This is the most probable state. The best distribution depends on whether these particles are classical particles, bosons or fermions. Why does the type of particle matter? It matters because of the properties of the *type*. The difference between fermions and the others is perhaps the most obvious: fermions cannot occupy the same state. So, if a given box (volume in phase space) has 40 states, only 40 fermions can be in that box while more than 40 bosons or classical particles can be in that box. The counting of number of particle arrangements is a bit different for classical and quantum particles (either bosons or fermions) because classical particles are identical, but distinguishable, while quantum particles are indistinguishable. For boxes with 40 states, each, then, we must maximize the number of possible arrangements of particles in the 40 states among the boxes available.

This animation shows a systems of 11 boxes, labeled 0 to 10, with energy equivalent to the box label. Each box has 40 states and there are 100 particles to distribute among the boxes. Select an average energy per particle and a distribution ("B-E" is Bose-Einstein for bosons, "F-D" is Fermi-Dirac for fermions, and "M-B" is Maxwell-Boltzmann for a classical system or collection of distinguishable particles). The black bar is a ratio of the probability of the particles in the configuration shown compared with the probability of particles in the most probable configuration for that energy and type of particle. In order for the total number of particles and the total energy to remain constant, decreasing particles from a box puts one particle in a higher energy box and one in a lower energy box. Similarly, increasing particles in a box requires one particle to move from each neighboring box. To begin, decrease the particles in the box they all initially begin in. Adjust the particle distribution to increase the total probability for the distribution of particles in the boxes.

Notice that your probability ratio increases as you spread the particles out more. At an average energy per particle of 5, what looks like it will be the best distribution? How do the bosons compare with the fermions and the classical particles? What about for an energy per particle of 1?

Mathematically, the count of the number of particle arrangements is called the multiplicity of states, W, and is given by the following for the three different cases:[1]

$$W_{\text{B-E}} = \prod \frac{(n_i + g_i - 1)!}{n_i!(g_i - 1)!} \quad W_{\text{F-D}} = \prod \frac{g_i!}{n_i!(g_i - n_i)!} \quad \text{and} \quad W_{\text{M-B}} = N! \prod \frac{g_i^{n_i}}{n_i!},$$
$$(15.1)$$

where g_i is the number of states in the ith box, n_i is the number of particles in the ith box, N is the total number of particles and $\prod$ signifies that we must multiply each ith term together from $i = 0$ to $i = N$. So, for this animation, $g_i = 40$ for all boxes and $N = 100$. Note that the multiplicity, W is related to entropy, S: $S = k_B \ln W$, where k_B is the Boltzmann constant.

Taking the above expressions for the possible arrangements of particles and maximizing them, we find the following particle distributions:[2]

$$n_{\text{B-E}\,i} = \frac{g_i}{e^{\alpha + \beta \varepsilon_i} - 1}, \quad n_{\text{F-D}\,i} = \frac{g_i}{e^{\alpha + \beta \varepsilon_i} + 1}, \quad \text{and} \quad n_{\text{M-B}\,i} = \frac{g_i}{e^{\alpha + \beta \varepsilon_i}}, \quad (15.2)$$

where α and β are set so that $\sum n_i = N$ and $\sum n_i \varepsilon_i = E$.

Solving these for our particular set up (11 boxes, 100 particles, and 40 states in each box), you can click below to see the ideal distribution for the energy/particle of interest (change the energy/particle above).

How do the ideal distributions compare at different values of average energy per particle? Notice that the biggest differences occur for the lowest energies. Why?

---

[1]See Kittel and Kroemer, *Thermal Physics*, 2nd edition (1980), for details of calculation of probabilities of distributions: fermions, pp. 10-15, bosons, p. 25 and classical systems, pp. 75-76.

[2]Actually, we generally maximize $\ln W$ so we can use Stirling's approximation to handle factorials of large numbers. We then set $d(\ln W) = 0$. To ensure that the number of particles stays fixed, as does the energy, we add the terms $\alpha \sum dn_i = 0$ and $\beta \sum \varepsilon_i dn_i = 0$ to $d(\ln W) = 0$ and solve for n_i that maximizes $\ln W$.

15.4   EXPLORING CLASSICAL, BOSE-EINSTEIN, AND FERMI-DIRAC STATISTICS

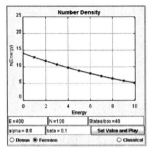

FIGURE 15.4: Fermion number density vs. energy for a system of particles.

This animation allows you to change the total number of particles and the total energy of a particular system of 11 boxes, each with an energy of 0 to 10. You can also change the number of states in a box. After making a change, you must push return (so the input box is no longer yellow) and you must push the "Set Value and Play" button. Each box represents a region in phase space with the same energy as set up in Section 15.3 and shown below:

FIGURE 15.5: Configuration for a system of particles.

To match the distributions you found in Section 15.3, keep number of states/box equal to 40 and number of particles equal to 100.

The animation calculates the values of α and β in the following expressions so that the total energy is constant and the total number of particles is constant:

$$n_{\text{B-E}\,i} = \frac{g_i}{e^{\alpha+\beta\varepsilon_i} - 1}, \qquad n_{\text{F-D}\,i} = \frac{g_i}{e^{\alpha+\beta\varepsilon_i} + 1}, \qquad \text{and} \qquad n_{\text{M-B}\,i} = \frac{g_i}{e^{\alpha+\beta\varepsilon_i}}, \quad (15.3)$$

where n_i is the number of particles in region in phase space, ε_i is the energy of that region (or box), and g is the number of states per box, also called a density of states.[3] The constants α and β must be set so that $\sum n_i = N$ and $\sum n_i \varepsilon_i = E$.

(a) Start with $E = 200$ and $N = 100$. The average energy per particle is 2. How does the plot of n_i vary from distribution to distribution? Explain the differences. Increase the number of particles to 200.

---

[3]Here we use a constant value for the number of states per box, g. This is valid in the one-dimensional case (assuming the degeneracy does not change with energy or location in phase space). Section 15.5 develops an expression for g in terms of momentum, p, for continuous distributions, $g(p)dp$, which can then be used to model systems we observe. Finally, since the energy increases linearly with box number, this is a model of a collection of one-dimensional harmonic oscillators.

(b) This animation does not have any protection against dropping the energy too low for a Fermi-Dirac distribution to exist. For example, with $E = 100$ and $N = 100$, drop the number of states, g, to 20. This means that only 20 particles can go into each state. If the particles sit in the lowest energy states possible, what is the lowest possible total energy? What is the last box that is full? and what is the corresponding energy of that box ($\varepsilon = 0$, $\varepsilon = 1$, $\varepsilon = 2$)? The last box that is filled (when all the lower energy boxes are full and none of the higher energy boxes have particles) is called the Fermi energy.

(c) If the total number of particles, N, is much less than the number of states, g, in each box, how do the distributions compare? Try several values of total energy and explain (be sure to try values of average energy per particle below 1).

(d) Keeping the number of particles, N, fixed, and the number of states fixed, change the total energy, E. How does β change?

What is β? If we start with the fundamental definition of temperature, we find that $\beta = 1/k_B T$ for all distributions as follows: For constant volume and particle number, temperature is defined by the relationship $dE = TdS$ where S is the entropy and E is the internal energy. Finding the distribution of n_i as a function of ε_i (as in the animation) requires maximizing $S = k_B \ln W$, where W is the count of the total number of states (see Section 15.3). In other words, we find the value of n_i that gives $d(\ln W) = \sum(\partial \ln W / \partial n_i)dn_i = 0$. There are, however, two other conditions to meet (which determine the values of α and β): keeping the number of particles and the total energy fixed. In equation form, this means $\sum n_i = N$ and $\sum n_i \varepsilon_i = E$ or that $\alpha \sum dn_i = 0$ and $\beta \sum \varepsilon_i dn_i = 0$. Thus, to find α and β (numerically in this animation), we solve $\sum(d \ln W / dn_i)dn_i - \alpha \sum dn_i - \beta \sum \varepsilon_i dn_i = 0$. However, built into this equation is the relationship $\sum(d \ln W / dn_i)dn_i = \beta \sum \varepsilon_i dn_i$, or

$$dS/k_B = \beta dE \quad \text{and thus} \quad \beta = 1/k_B T. \tag{15.4}$$

What about α? Section 15.5 shows how it depends on the type of system.

15.5 STATISTICS OF AN IDEAL GAS, A BLACKBODY, AND A FREE ELECTRON GAS

In this animation, you can choose a specific idealized system by pushing the radio buttons. You can change the value of $k_B T$ with the slider and the value of the constant in the equation described below for the specific system (labeled as C). You can also plot the distribution function $n(\varepsilon)$ and also $n(\varepsilon)/g(\varepsilon)$, which is the occupancy, $f(\varepsilon)$, and should match the shape of the distributions found in Section 15.4.

Different assemblies of particles are described by the relevant distribution function (number of particles with a given energy):

	Discrete	Continuous
B-E	$n_i = g_i/(e^{\alpha + \beta \varepsilon_i} - 1)$	$n(\varepsilon)d\varepsilon = g(\varepsilon)d\varepsilon/(e^{\alpha + \beta \varepsilon} - 1)$
F-D	$n_i = g_i/(e^{\alpha + \beta \varepsilon_i} + 1)$	$n(\varepsilon)d\varepsilon = g(\varepsilon)d\varepsilon/(e^{\alpha + \beta \varepsilon} + 1)$
M-B	$n_i = g_i/e^{\alpha + \beta \varepsilon_i}$	$n(\varepsilon)d\varepsilon = g(\varepsilon)d\varepsilon/e^{\alpha + \beta \varepsilon}$

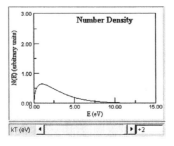

FIGURE 15.6: The number density vs. energy for an ideal gas.

where n_i is the number of particles in region in phase space (a given box, see Section 15.3), ε_i is the energy of that region (or box) and g_i is the number of states associated with a given region in phase space (a given box), also called a density of states. The constants α and β must be set so that $\sum n_i = N$ (or $\int dn = N$) and $\sum n_i \varepsilon_i = E$ (or $\int \varepsilon n(\varepsilon) d\varepsilon = E$). From the definition of temperature, $\beta = 1/k_B T$ for all distributions (see Section 15.4). However, the value of α varies by distribution, so we will consider several specific applications.[4]

In order to determine which distribution to use, we must remember that the fundamental difference between quantum-mechanical distributions and the classical distribution is whether or not the particles are distinguishable. In both cases, the particles are identical, but, for the quantum distributions, they are countable, but indistinguishable. One way to determine whether we can consider particles distinguishable or not is whether or not the wave function for each particle can be considered separately from the other particle or not. For example, the wave function for the two electrons in a helium atom cannot be written as two individual wave functions that do not overlap significantly, but the wave function for the electrons in two neighboring hydrogen atoms (in a diffuse gas) can be considered independently of each other (because the wave function between the two essentially goes to zero and they no longer need to be considered entangled states).

Ideal gas is a model of hard spheres that collide into each other and do not otherwise interact with each other. For a diffuse gas, the molecules are considered distinguishable so we can use the Maxwell-Boltzmann (classical) distribution. Furthermore, the number of particles is generally quite large so that energy quantization is not noticeable (at normal temperatures). Therefore, we can use a continuous distribution.

It is easier to determine $g(p)dp$, the number of states with momentum between p and dp, and then using $\varepsilon = p^2/2m$ rewrite $g(p)dp$ as $g(\varepsilon)d\varepsilon$:

$$g(p)dp = \int\int\int\int\int\int dx\,dy\,dz\,dp_x\,dp_y\,dp_z/h^3 = V4\pi p^2 dp/h^3 \ , \quad (15.5)$$

[4]Development follows A. Beiser, *Modern Physics*, 6th ed, McGraw-Hill, 2003, pp. 296-331 and J. R. Taylor, C. D. Zapfiratos and M. A. Dubson, *Modern Physics*, 2nd ed, Prentice Hall, 2004, pp. 502-526.

where V is the volume integral of $dx\,dy\,dz$ and $4\pi p^2 dp$ is due to changing $dp_x\,dp_y\,dp_z$ from Cartesian to spherical coordinates and integrating. We find α by requiring that $\int_0^\infty n(\varepsilon)d\varepsilon = N$:

$$n(\varepsilon)d\varepsilon = (4\pi V/h^3)m^{3/2}\varepsilon^{1/2}e^{-\alpha}e^{-\varepsilon/k_BT}d\varepsilon\,, \qquad (15.6)$$

which, upon integration, gives $e^{-\alpha} = (Nh^3/V)(2\pi m k_B T)^{-3/2}$ so that finally

$$n(\varepsilon)d\varepsilon/N = 2\pi/(\pi k_B T)^{3/2}\varepsilon^{1/2}e^{-e/k_BT}d\varepsilon \qquad (15.7)$$

Click on the "Ideal Gas" button. You can change the value of k_BT with the slider and the value of $(Nh^3/V)(2\pi m)^{-3/2}$ (labeled as C). How does the distribution change with C? Explain.

Blackbody Radiation (see also Section 4.1) is the radiation from a perfect emitter (absorber) and is due to photons, which are indistinguishable particles of spin 1 (and therefore are bosons). This means that we will use Bose-Einstein statistics. The energy levels of the photons are quantized, but there are so many different energy levels, we can use the continuous distribution function. The number of states, $g(p)dp$ in a given region of phase space (with an energy between p and $p + dp$) is double that of the ideal gas because there are two polarization states. Using momentum as $p = \varepsilon/c$ we can find $g(\varepsilon)d\varepsilon$ as below:

$$g(\varepsilon)d\varepsilon = \frac{8V\pi\varepsilon^2 d\varepsilon}{(hc)^3}\,. \qquad (15.8)$$

Furthermore, the total number of photons, N, does not have to remain constant. This means that we can set $\alpha = 0$. This gives

$$n(\varepsilon)d(\varepsilon) = \frac{8V\pi}{h^3c^3}\frac{\varepsilon^2 d\varepsilon}{(e^{\varepsilon/k_BT}-1)}\,, \qquad (15.9)$$

Click on the distribution for blackbody radiation (since $\alpha = 0$, changing C in the animation does not change anything). Notice that the number of particles (area under the $n(\varepsilon)$ curve) is not constant. Look at $f(\varepsilon)$. What happens when the temperature goes to zero?

Free Electron Gas is used to describe the energy distribution of electrons in metals. Electrons in the conduction bands of metals that are essentially free to move throughout the material, so we describe this as a free electron *gas*. Since this is a gas of electrons, however, we must use Fermi-Dirac statistics. Again, although the energies are quantized, there are many energy levels in the energy band and so we can use a continuous distribution. As before, we need to find $g(\varepsilon)d\varepsilon$ and α. Again, it is easier to use $g(p)dp$ and it is twice the value of the ideal gas expression because there are two spin states. Using $\varepsilon = p^2/2m$ (and thus, $dp = (m/2\varepsilon)^{1/2}d\varepsilon$), we have

$$g(\varepsilon)d\varepsilon = \frac{8V\pi(2m^3)^{1/2}}{h^3}\varepsilon^{1/2}d\varepsilon\,, \qquad (15.10)$$

Now, we need to find α. It turns out that to be useful to describe the temperature dependence of α as $\alpha = -\varepsilon_F/k_B T$, where ε_F is the Fermi energy of a metal. So,

$$n(\varepsilon)d\varepsilon = \frac{8V\pi(2m^3)^{1/2}}{h^3} \frac{\varepsilon^{1/2}d\varepsilon}{\left(e^{-(\varepsilon_F+\varepsilon)/k_B T} + 1\right)} . \qquad (15.11)$$

Notice what happens when $T = 0$. When $\varepsilon > \varepsilon_F$, $n(\varepsilon)d\varepsilon = 0$. When $\varepsilon_F > \varepsilon$, $n(\varepsilon)d\varepsilon = g(\varepsilon)d\varepsilon$. In words, then, the states below the Fermi energy have particles in them (at $T = 0$), but the states above the Fermi energy do not. Notice that for blackbody radiation above, when $T = 0$, $n(\varepsilon)d\varepsilon = 0$ except for the state $\varepsilon = 0$. For a boson, at $T = 0$ all the particles are in the lowest energy region in phase space, but for a fermion many regions in phase space are occupied because the Pauli exclusion principle does not allow particles to share states. This limit allows for the calculation of the Fermi energy at $T = 0$: $\int_0^{\varepsilon_F} n(\varepsilon)d\varepsilon = \int_0^{\varepsilon_F} g(\varepsilon)d\varepsilon = N$ and the limits of integration are from 0 to ε_F because $n(\varepsilon)d\varepsilon = 0$ for $\varepsilon > \varepsilon_F$. Solving, then, we find

$$\varepsilon_F = \frac{h^2}{2m}\left(\frac{3N}{8\pi V}\right)^{2/3} . \qquad (15.12)$$

Click on the "Free Electron Gas" distribution. For a metal with $\varepsilon_F = 5$ eV, find what happens to the distribution as $T = 0$? What about the occupancy? At room temperature, how does β compare with the Fermi level? This particular analysis of metals as a free electron gas is only valid at low temperatures (so electrons are not excited to higher energy bands) so we keep the value of $k_B T < 0.2$ eV in the animation.

15.6 EXPLORING THE EQUIPARTITION OF ENERGY

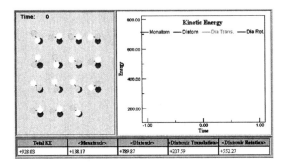

FIGURE 15.7: A model of a gas illustrating the equipartition of energy.

The equipartition of energy theorem says that the energy of an atom or particle is, on average, equally distributed between the different modes (different degrees of freedom) available. The way to count the modes, or degrees of freedom, is to count the number of quadratic terms in the energy expression. For example, for a monatomic gas (without any external forces), the energy of a particle is given as $(1/2)mv_x^2 + (1/2)mv_y^2 + (1/2)mv_z^2$. There are three terms that are quadratic. However, for a particle in a three-dimensional simple harmonic potential well, there are six different modes (degrees of freedom). The energy per particle has an average value of $(f/2)k_BT$, where f is the number of degrees of freedom, k_B is the Boltzmann constant and T is the temperature.

You can verify this result by taking the distribution function for a monatomic ideal gas and finding the total energy:[5]

$$n(\varepsilon)d\varepsilon = \frac{2\pi N}{(\pi kBT)^{3/2}} \varepsilon^{1/2} e^{-\varepsilon/k_BT} d\varepsilon ,\qquad (15.13)$$

where $\int n(\varepsilon)d\varepsilon = E$ and $E = (3/2)Nk_BT$. For a harmonic oscillator, the distribution is different, but in the end it gives you $E = 3k_BT$.

(a) In this animation of a monatomic gas in a box, why do the particles only have 2 degrees of freedom? The table shows the total kinetic energy of all particles in the box, as well as the average kinetic energies of particles in the box (the animation averages over a 10-s period, so you need to wait 10 s to read the averages).

(b) Record the total energy. What is the energy per particle? If the energy is given in joules/k_B, what is the temperature inside the box?

Try this animation of a diatomic gas with 20 particles. Notice that the graph shows the total kinetic energy of the diatomic particles and the kinetic energies of translation (motion in x and y directions) and rotation.

(c) Why is the translational kinetic energy, on average, about two times the rotational kinetic energy?

(d) From the total energy, what is the energy per particle? If the energy is given in joules/k_B, what is the temperature in the box? Hint: Remember that $\langle \text{energy} \rangle / \text{particle} = (f/2)k_BT$ and in this case, $f = 3$ (Why?).

Now, try a mixture of 20 monatomic particles and 20 diatomic particles.

(e) Why is the temperature of the gas in the box a single value (not one value for atoms and another for molecules)?

(f) After waiting at least 10 s, compare the average values of the kinetic energies. What value is the average monatomic kinetic energy close to? Why should those two values, averaged over a long period of time, be equal to each other and greater than the rotational kinetic energy of the diatomic particles?

(g) From the total energy (given in joules/kB), what is the temperature? Hint: Explain why the total energy should be equal to $(2/2)20k_BT + (3/2)20k_BT$.

---

[5]This is a classical result because we are using the classical distribution (Section 15.5). If the temperature is very low and the density high, we will need to use a quantum mechanical distribution and these results do not hold (see Section 15.7 for an example of low temperature calculations in a solid).

(h) In this animation, if a mixture has 15 atoms, how many diatomic particles should it have so that the average kinetic energies of both particles are the same? Try setting the number of monatomic particles and diatomic particles to check your answer.

15.7 SPECIFIC HEAT OF SOLIDS

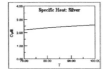

FIGURE 15.8: Specific heat vs. temperature for silver.

The measure of the change in temperature with the change in internal energy (heating an object up, for example) is called the specific heat (at a constant volume), c_V, and is given by

$$c_V = \frac{dE}{dT} . \tag{15.14}$$

To find the specific heat of a solid, then, we simply need to find an expression of the total energy as a function of temperature.

One way is to imagine a solid as composed of atoms in a lattice, each individually sitting in its own three dimensional simple harmonic well in which it vibrates with quantized energies of vibration. In this picture, we can use Maxwell-Boltzmann statistics (and the corresponding distributions) and find the following expression for the specific heat:

$$c_V = 3R \left(\frac{\hbar\omega}{k_B T}\right)^2 \frac{e^{\hbar\omega/k_B T}}{(e^{\hbar\omega/k_B T} - 1)^2} , \tag{15.15}$$

where we pick the value of $\hbar\omega$ to match the experimental data. Note that for high temperatures ($k_B T \gg \hbar\omega$), the value of the specific heat is consistent with the equipartition of energy theorem: $c_V = (f/2)R = 3R$ (since $f = 6$, why?).

At low temperatures, however, the previous model fails. In this case, the coupling between the oscillators cannot be ignored. The vibration frequency of one atom does have an impact on its neighbor. So, instead of treating the oscillators individually, we must treat them as a group of points on an elastic spring that oscillates up and down. These oscillations of the group as a whole create elastic standing waves in the solid called phonons. Phonon statistics are then treated as photons in blackbody radiation except that there is a limit to the phonon energy (the vibration wavelength and thus, frequency, is limited by the spacing between atoms). This gives a specific heat as follows:[6]

$$c_V = -9R \left(\frac{\Theta}{T}\right) \frac{1}{e^{\Theta/T} - 1} + 36R \left(\frac{T}{\Theta}\right)^3 \int_0^{\Theta/T} \frac{x^3}{(e^x - 1)} dx , \qquad (15.16)$$

where Θ is the Debye temperature and can be measured independently. At low temperatures ($T << \Theta$), the integral can be evaluated analytically giving

$$c_V = \frac{12\pi^4 R}{5} \left(\frac{T}{\Theta}\right)^3 . \qquad (15.17)$$

Compare a plot of this expression with the classically derived version for silver from 0 to 30 K. The red plot is the Debye model at low temperatures.

At high temperatures, both expressions give the same result. So, in general we use the following

$$c_V = 3R \left(\frac{A}{T}\right)^2 \frac{e^{A/T}}{(e^{A/T} - 1)^2} \quad \text{and} \quad c_V = \frac{12\pi^4 R}{5} \left(\frac{T}{\Theta}\right)^3 , \qquad (15.18)$$

for high temperatures (derived classically) and for low ($T < 10 - 20\%$ of Θ) temperatures (derived quantum mechanically), respectively. What determines a high and low temperature of a material? It depends on Θ, the Debye temperature. In order to compare the two functions, for materials of interest here, $A = 0.65\Theta$. Looking at the data in the temperature ranges given, what is considered *low temperature* for iron? Does it have a lower or higher Debye temperature than silver?

PROBLEMS

15.1. This animation shows a systems of 11 boxes, labeled 0 to 10, of energy equivalent to the box label. Each box has 40 states and there are 100 particles to distribute among the boxes. (B-E is Bose-Einstein for bosons, F-D is Fermi-Dirac for fermions, and M-B is Maxwell-Boltzmann for a classical system). The equilibrium distribution is shown. What is the average energy per particle?

---

[6]See R. Eisberg and R. Resnick, *Quantum Physics*, Wiley, 1974, pp. 421-425 for a detailed analysis of both the classical and proper quantum-mechanical derivation of the specific heat of solids.

15.2. This animation allows you to change the total number of particles and the total energy of a particular system of 11 boxes, each with an energy of 0 to 10. You can also change the number of states in a box currently set to 40. To make a change, you must push return (so the input box is no longer yellow) and you must push the "Set Value and Run" button. The animation calculates the values of α and β in the following expressions so that the total energy is constant and the total number of particles is constant ($\beta = 1/k_B T$ for all distributions)

$$n_{\text{B-E}\,i} = \frac{g_i}{e^{\alpha+\beta\epsilon_i} - 1} \,, \quad n_{\text{F-D}\,i} = \frac{g_i}{e^{\alpha+\beta\epsilon_i} + 1} \,, \quad \text{and } n_{\text{M-B}\,i} = \frac{g_i}{e^{\alpha+\beta\epsilon_i}} \,, \quad (15.19)$$

where n_i is the number of particles in region in phase space, ε_i is the energy of that region (or box) and g is the number of states per box, also called a density of states.

(a) For situations where the average energy per particle is the same, how does the temperature of Fermi-Dirac distribution compare with a classical system? Specifically, for an average energy per particle of 5, how do the temperatures compare? What about for an average energy per particle of 1?

(b) The average energy per particle of electrons in a metal (Fermi gas) are close to the Fermi level at very low temperatures (T near zero). If the electrons in a metal with a Fermi energy of 5 eV behaved classically (like ideal gas particles), what temperature would be necessary for them to have the same average energy per particle as they actually have at T near zero?

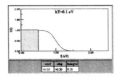

15.3. The graph shown represents the probability of occupation for an electron in a substance with a uniform density of states (unlike a metal where the density of states is proportional to $E^{1/2}$). This function can be plotted for several different temperatures. The integral shown is normalized and represents the probability of finding electrons of a certain energy range. What percentage of the electrons are within 10% of the Fermi energy at the various temperatures? Explain.

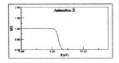

15.4. The graphs shown represent the probability of occupation electrons in a metal as a function of energy.

(a) Rank the animations by Fermi energy.

(b) Rank the animations by temperature.

15.5. The graph shown represents the density of occupied states for electrons in a metal depending on energy. This function can be plotted for several different temperatures. What value of kT corresponds to 7% of the electrons having energy greater than the Fermi energy?

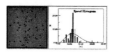

15.6. In this animation $N = nR$ (i.e., $k_B = 1$). This, then, gives the ideal gas law as $PV = NT$. The average values shown, $\langle\ \rangle$, are calculated over intervals of one time unit so that the average rate change in momentum is equal to the pressure times the area, A (where $A = 1$). This animation shows the distribution of speeds in an ideal gas based on the Maxwell-Boltzmann distribution as shown by the smooth black curve on the graph for a given temperature:

$$n(v)dv/N = (2/\pi)^{1/2}(m/k_BT)^{3/2}v^2e^{-mv^2/2k_BT} . \qquad (15.20)$$

What happens to the distribution as you increase the energy (temperature)? Since there is a speed distribution, when we talk about a characteristic speed of a gas particle at a particular temperature, we use one of three characteristic speeds:

- Average speed ($\langle v \rangle = \int vn(v)dv$).
- Most probable speed (find maximum of $n(v)$).
- Root-mean-square (rms) speed ($\langle v^2 \rangle^{1/2} = [\int v^2n(v)dv]^{1/2}$).

Find an expression for each (in terms of m and k_BT). Identify which peak is which characteristic speed.

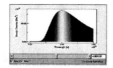

15.7. The graph represents energy density per wavelength emitted from a hot object versus the wavelength of that light. You can change the temperature of the object (with the slider) and the value of the maximum wavelength.

Starting with the distribution of particles as a function of energy for blackbody radiation (see Section 15.4) below: $n(\varepsilon)d\varepsilon = (8V\pi/h^3c^3)e^2d\varepsilon/(e^{\varepsilon/k_BT} - 1)$. Rewrite the expression as a function of $\lambda n(\lambda)d\lambda$, where $\varepsilon = hc/\lambda$. The plot above is a plot of the energy per volume as a function of λ, called an energy density given by $\varepsilon(\lambda)/V = \varepsilon n(\lambda)d\lambda/V$. The peak of the curve is the wavelength where $d(\varepsilon(\lambda)/V)/d\lambda = 0$.

Find an expression for the wavelength where the energy density is a maximum (for a given temperature) and verify it using the animation.

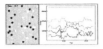

15.8. The gas in the animation has 20 monatomic particles and 10 diatomic particles. Which of the lines on the graph corresponds to the total kinetic energy, the total kinetic energy of the monatomic particles, the total kinetic energy of the diatomic particles, and the translational and rotational kinetic energy of the diatomic particles? Explain.

15.9. The graphs show the specific heat as a function of temperature for three solids. You can choose three temperature ranges. Rank the solids by Debye temperature.

15.10. The spectrum above is an approximation to the P and R branches of the rotational-vibrational absorption spectrum of CO. This is a spectrum for the case of a transition from the vibrational ground state to the first excited vibrational state. Each vibrational energy level, however, contains a number of rotational energy levels because the energy between rotational energy levels is in the 10^{-4} eV range while the energy difference between vibrational levels is 10^{-1} eV. So, as the molecule goes from ω_0 to ω_1 (ground to first excited vibrational states), there is also a transition between rotational states (governed by the selection rule $\Delta J = \pm 1$ where J is the angular momentum quantum number for the molecule). The P-branch are transitions where $\Delta J = -1$ and the R branch, $\Delta J = +1$. The intensity depends on the initial population of the states. The rotational energy of the molecule is given classically by $E = L^2/2I$ (kinetic energy of rotation where I is the moment of inertia of the molecule and L is the classical angular momentum). The rotational levels are quantized and we will use J as the angular momentum quantum number so that the energy of rotation is given by $\varepsilon_J = J(J+1)\hbar^2/2I$.
  (a) Why is the density of states, $g(\varepsilon)$ given by $g(\varepsilon) = 2J + 1$?
  (b) Each of the molecules is considered indistinguishable, so we can use Maxwell-Boltzmann statistics: $n_i = g_i/e^{\alpha + \beta \varepsilon_i}$. Let $e^{-\alpha} = n_0$. Write an expression for n_J, the number of particles with a given value of J.
  (c) This distribution has a maximum at a value of J that depends on the moment of inertia I and temperature T. Find an expression for the value of J as a function of I and T that corresponds to the maximum value of n_J. From the spectrum, what value of J corresponds to the most intense peak (and thus had the highest number of particles in it prior to absorption of energy).
  (d) Determine, finally, the value of the moment of inertia of CO and compare it with the value you get simply by noting the energy difference between the different peaks.

Bibliography

1. D. Styer, "Quantum Mechanics: See it Now," AAPT Kissimmee, FL January, 2000 and http://www.oberlin.edu/physics/dstyer/TeachQM/see.html.

2. D. Styer, "Common Misconceptions Regarding Quantum Mechanics," *Am. J. Phys.* **64**, 31-34 (1996).

3. R. W. Robinett, *Quantum Mechanics: Classical Results, Modern Systems, and Visualized Examples*, Oxford, New York, 1997.

4. E. Cataloglu and R. Robinett, "Testing the Development of Student Conceptual and Visualization Understanding in Quantum Mechanics through the Undergraduate Career," *Am. J. Phys.* **70**, 238-251 (2002).

5. D. Zollman, *et al.*, "Research on Teaching and Learning of Quantum Mechanics," *Papers Presented at the National Association for Research in Science Teaching* (1999).

6. C. Singh, "Student Understanding of Quantum Mechanics," *Am. J. Phys.* **69**, 885-895 (2001).

7. R. Müller and H. Wiesner, "Teaching Quantum Mechanics on the Introductory Level," *Am. J. Phys.* **70**, 200-209 (2002).

8. L. Bao and E. Redish, "Understanding Probabilistic Interpretations of Physical Systems: A Prerequisite to Learning Quantum Physics," *Am. J. Phys.* **70**, 210-217 (2002).

9. D. Zollman, N. S. Rebello, and K. Hogg, "Quantum Mechanics for Everyone: Hands-on Activities Integrated with Technology," *Am. J. Phys.* **70**, 252-259 (2002).

10. S. Brandt and H. Dahmen, *The Picture Book of Quantum Mechanics*, Springer-Verlag, New York, 2001.

11. J. Hiller, I. Johnston, D. Styer, *Quantum Mechanics Simulations, Consortium for Undergraduate Physics Software*, John Wiley and Sons, New York, 1995.

12. B. Thaller, *Visual Quantum Mechanics*, Springer-Verlag, New York, 2000.

13. M. Joffre, *Quantum Mechanics CD-ROM* in J. Basdevant and J. Dalibard, *Quantum Mechanics*, Springer-Verlag, Berlin, 2002.

14. A. Goldberg, H. M. Schey, and J. L. Schwartz, "Computer-generated Motion Pictures of One-dimensional Quantum-mechanical Transmission and Reflection Phenomena," *Am. J. Phys.* **35**, 177-186 (1967).

15. M. Andrews, "Wave Packets Bouncing Off of Walls," *Am. J. Phys.* **66** 252-254 (1998).

16. M. A. Doncheski and R. W. Robinett, "Anatomy of a Quantum 'Bounce,'" *Eur. J. Phys.* **20**, 29-37 (1999).

17. M. Belloni, M. A. Doncheski, and R. W. Robinett, "Exact Results for 'Bouncing' Gaussian Wave Packets," *Phys. Scr.* **71**, 136-140 (2005).

18. J. J. Sakurai, *Advanced Quantum Mechanics*, Addison-Wesley (1967).

19. R. E. Scherr, P. S. Shaffer, and S. Vokos, "The Challenge of Changing Deeply Held Student Beliefs about the Relativity of Simultaneity," *Am. J. Phys.* **70**, 1238 (2002).

20. R. E. Scherr, P. S. Shaffer, and S. Vokos, "Student Understanding of Time in Special Relativity: Simultaneity and Reference Frames," *Phys. Educ. Res., Am. J. Phys. Suppl.* **69**, S24 (2001).

21. K. Krane, *Modern Physics*, 2nd edition, John Wiley and Sons (1996).

22. P. A. Tipler and R. A. Llewellyn, *Modern Physics*, W. H. Freeman and Company (1999).

23. J. R. Taylor, C. H. Zafiratos, and M. A. Dubson, *Modern Physics for Scientists and Engineers*, Prentice Hall (2004).

24. S. Thornton and A. Rex, *Modern Physics for Scientists and Engineers*, 2nd ed, Brooks/Cole (2002).

25. W. E. Lamb, Jr. and M. O. Scully, "The Photoelectric Effect without Photons," in *Polarisation, Matièrer et Rayonnement*, Presses University de France (1969).

26. G. Greenstein and A. G. Zajonc, *The Quantum Challenge*, Jones and Bartlett (1997).

27. J. J. Thorn, M. S. Neel, V. W. Donato, G. S. Bergreen, R. E. Davies, and M. Beck, "Observing the Quantum Behavior of Light in an Undergraduate Laboratory," *Am. J. Phys.* **72** 1210-1219 (2004).

28. D. F. Styer, *et al.*, "Nine Formulations of Quantum Mechanics," *Am. J. Phys.* **70**, 288-297 (2002).

29. M. Belloni, M. A. Doncheski, and R. W. Robinett, "Zero-curvature solutions of the one-dimensional Schrödinger equation," to appear in *Phys. Scr.* 2005.

30. L. P. Gilbert, M. Belloni, M. A. Doncheski, and R. W. Robinett, "More on the Asymmetric Infinite Square Well: Energy Eigenstates with Zero Curvature," to appear in *Eur. J. Phys.* 2005.

31. L. P. Gilbert, M. Belloni, M. A. Doncheski, and R. W. Robinett, "Piecewise Zero-curvature Solutions of the One-Dimensional Schrödinger Equation," in preparation.

32. R. W. Robinett, "Quantum Wave Packet Revivals," talk given at the 128th AAPT National Meeting, Miami Beach, FL, Jan. 24-28 (2004).

33. R. Shankar, *Principles of Quantum Mechanics*, Plenum Press (1994).

34. M. Morrison, *Understanding Quantum Physics: A Users Manual*, Prentice Hall, Upper Saddle River, NJ, 1990.

35. M. Bowen and J. Coster, "Infinite Square Well: A Common Mistake," *Am. J. Phys.* **49**, 80-81 (1980)

36. R. C. Sapp, "Ground State of the Particle in a Box," *Am. J. Phys.* **50**, 1152-1153 (1982)

37. L. Yinji and H. Xianhuai, "A Particle Ground State in the Infinite Square Well," *Am. J. Phys.* **54**, 738 (1986).

38. C. Dean, "Simple Schrödinger Wave Functions Which Simulate Classical Radiating Systems," *Am. J. Phys.* **27**, 161-163 (1959).

39. R. W. Robinett, "Quantum Wave Packet Revivals," *Phys. Rep.* **392**, 1-119 (2004).

40. R. Bluhm, V. A. Kostelecký, and J. Porter, "The Evolution and Revival Structure of Localized Quantum Wave Packets," *Am. J. Phys.* **64**, 944-953 (1996).

41. I. Sh. Averbukh and N. F. Perelman, "Fractional Revivals: Universality in the Long-term Evolution of Quantum Wave Packets Beyond the Correspondence Principle Dynamics," *Phys. Lett.* **A139**, 449-453 (1989)

42. D. L. Aronstein and C. R. Stroud, Jr., "Fractional Wave-function Revivals in the Infinite Square Well," *Phys. Rev. A* **55**, 4526-4537 (1997).

43. R. Liboff, *Introductory Quantum Mechanics*, Addison Wesley (2003).

44. F. Bloch, *Z. Physik*, **52** (1928).

45. M. A. Doncheski and R. W. Robinett, "Comparing classical and quantum probability distributions for an asymmetric well," *Eur. J. Phys.* **21**, 217–228 (2000).

46. A. Bonvalet, J. Nagle, V. Berger, A. Migus, J.-L. Martin, and M. Joffre, "Femtosecond Infrared Emission Resulting from Coherent Charge Oscillations in Quantum Wells," *Phys. Rev. Lett.* **76**, 4392-4395 (1996).

47. C. Kittel and H. Kroemer, *Thermal Physics*, 2nd ed, W. H. Freeman, 1980.

48. R. Eisberg and R. Resnick, *Quantum Physics*, Wiley, 1974.

License Agreement

Mario Belloni/Wolfgang Christian/Anne J. Cox
Physlet Quantum Physics CD-ROM
0-13-101970-8
©2006 Pearson Education, Inc.
Pearson Prentice Hall
Pearson Education, Inc.
Upper Saddle River, NJ 07458
All rights Reserved.
Pearson Prentice Hall™ is a trademark of Pearson Education, Inc.

YOU SHOULD CAREFULLY READ THE TERMS AND CONDITIONS BEFORE USING THE CD-ROM PACKAGE. USING THIS CD-ROM PACKAGE INDICATES YOUR ACCEPTANCE OF THESE TERMS AND CONDITIONS. Pearson Education, Inc. provides this program and licenses its use. You assume responsibility for the selection of the program to achieve your intended results, and for the installation, use, and results obtained from the program. This license extends only to use of the program in the United States or countries in which the program is marketed by authorized distributors.

LICENSE GRANT
You hereby accept a nonexclusive, nontransferable, permanent license to install and use the program ON A SINGLE COMPUTER at any given time. You may copy the program solely for backup or archival purposes in support of your use of the program on the single computer. You may not modify, translate, disassemble, decompile, or reverse engineer the program, in whole or in part.

TERM
The License is effective until terminated. Pearson Education, Inc. reserves the right to terminate this License automatically if any provision of the License is violated. You may terminate the License at any time. To terminate this License, you must return the program, including documentation, along with a written warranty stating that all copies in your possession have been returned or destroyed.

LIMITED WARRANTY
THE PROGRAM IS PROVIDED "AS IS" WITHOUT WARRANTY OF ANY KIND, EITHER EXPRESSED OR IMPLIED, INCLUDING, BUT NOT LIMITED TO, THE IMPLIED WARRANTIES OR MERCHANTABILITY AND FITNESS FOR A PARTICULAR PURPOSE. THE ENTIRE RISK AS TO THE QUALITY AND PERFORMANCE OF THE PROGRAM IS WITH YOU. SHOULD THE PROGRAM PROVE DEFECTIVE, YOU (AND NOT PEARSON EDUCATION, INC. OR ANY AUTHORIZED DEALER) ASSUME THE ENTIRE COST OF ALL NECESSARY SERVICING, REPAIR, OR CORRECTION. NO ORAL OR WRITTEN INFORMATION OR ADVICE GIVEN BY PEARSON EDUCATION, INC., ITS DEALERS, DISTRIBUTORS, OR AGENTS SHALL CREATE A WARRANTY OR INCREASE THE SCOPE OF THIS WARRANTY. SOME STATES DO NOT ALLOW THE EXCLUSION OF IMPLIED WARRANTIES, SO THE ABOVE EXCLUSION MAY NOT APPLY TO YOU. THIS WARRANTY GIVES YOU SPECIFIC LEGAL RIGHTS AND YOU MAY ALSO HAVE OTHER LEGAL RIGHTS THAT VARY FROM STATE TO STATE. Pearson Education, Inc. does not warrant that the functions contained in the program will meet your requirements or that the operation of the program will be uninterrupted or error-free. However, Pearson Education, Inc. warrants the CD-ROM(s) on which the program is furnished to be free from defects in material and workmanship under normal use for a period of ninety (90) days from the date of delivery to you as evidenced by a copy of your receipt. The program should not be relied on as the sole basis to solve a problem whose incorrect solution could result in injury to person or property. If the program is employed in such a manner, it is at the user's own risk and Pearson Education, Inc. explicitly disclaims all liability for such misuse.

LIMITATION OF REMEDIES
Pearson Education, Inc.'s entire liability and your exclusive remedy shall be:
1. the replacement of any CD-ROM not meeting Pearson Education, Inc.'s "LIMITED WARRANTY" and that is returned to Pearson Education, or

2. if Pearson Education is unable to deliver a replaceme CD-ROM that is free of defects in materials or workmansh you may terminate this agreement by returning the progra IN NO EVENT WILL PEARSON EDUCATION, INC. I LIABLE TO YOU FOR ANY DAMAGES, INCLUDING AN LOST PROFITS, LOST SAVINGS, OR OTHER INCIDENTA OR CONSEQUENTIAL DAMAGES ARISING OUT OF TH USE OR INABILITY TO USE SUCH PROGRAM EVE IF PEARSON EDUCATION, INC. OR AN AUTHORIZE DISTRIBUTOR HAS BEEN ADVISED OF THE POSSIBILIT OF SUCH DAMAGES, OR FOR ANY CLAIM BY ANY OTHE PARTY. SOME STATES DO NOT ALLOW FOR THE LIMIT TION OR EXCLUSION OF LIABILITY FOR INCIDENTAL C CONSEQUENTIAL DAMAGES, SO THE ABOVE LIMITATIC OR EXCLUSION MAY NOT APPLY TO YOU.

GENERAL
You may not sublicense, assign, or transfer the license of t program. Any attempt to sublicense, assign or transfer any of t rights, duties, or obligations hereunder is void. This Agreeme will be governed by the laws of the State of New York. Should y have any questions concerning this Agreement, you may conta Pearson Education, Inc. by writing to:

ESM Media Development
Higher Education Division
Pearson Education, Inc.
1 Lake Street
Upper Saddle River, NJ 07458

Should you have any questions concerning technical support, y may write to:

New Media Production
Higher Education Division
Pearson Education, Inc.
1 Lake Street
Upper Saddle River, NJ 07458

YOU ACKNOWLEDGE THAT YOU HAVE READ TH AGREEMENT, UNDERSTAND IT, AND AGREE TO I BOUND BY ITS TERMS AND CONDITIONS. YOU FURTHI AGREE THAT IT IS THE COMPLETE AND EXCLUSIV STATEMENT OF THE AGREEMENT BETWEEN US THA SUPERSEDES ANY PROPOSAL OR PRIOR AGREEMEN ORAL OR WRITTEN, AND ANY OTHER COMMUNICATIOI BETWEEN US RELATING TO THE SUBJECT MATTER C THIS AGREEMENT.

Program Instructions
To access the *Physlet Quantum Physics* curricular materi follow these steps:
-Insert the *Physlet Quantum Physics* CD into your CD-RC drive.
-Navigate to the *Physlet Quantum Physics* CD-ROM.
-Double-click the start.html file.

Minimum System Requirements:
Requires a browser supporting the Java 1.4.2 or higher (1. or higher, preferred) Virtual Machine and JavaScript to Ja communication.
32 MB or more of available RAM.
800x600 monitor resolution set to 16 bit color.
Mouse or other pointing device.
4x CD-ROM Drive.

-Windows
400 MHz processor
Windows 2000/XP
Internet Explorer 5.5, Mozilla 1.7, or Firefox 1.0 or higher
Sun Java plugin 1.4.2 or higher

-Macintosh
PowerPC G3, G4, or G5 processor
Mac OS X.3/X.4
Safari 1.2
Macintosh Java plugin 1.4.2 or higher
Navigate to each page as outlined in the Preface.